Finance for Teens: A Step-by-Step Smart Money Management Guide

Master the Skills to Budget, Save, Invest, Avoid Debt, and Build the Confidence to Live a Financially Secure and Independent Future

Jade Miles

CONTENTS

INTRODUCTION

With social media running the show these days, it's easy to believe everything we see online. You've probably scrolled past countless TikToks or Instagram Reels featuring teens with zero work experience and no degree, yet somehow they're raking in thousands from those seemingly easy "side hustles" overnight. Or those clips promising you can become a millionaire just by throwing your money into crypto. Meanwhile, you're over here struggling to make your allowance stretch or wondering why that summer job you busted your back for barely covers your needs, let alone your wants.

Here's the thing: Those "overnight success" stories? Yeah, they're not telling you the whole truth. A lot of these so-called influencers already had a head start—maybe their parents bankroll their jet-set lifestyle, or maybe they're making bank off sponsorship deals while trying to sell you a dream. Those flashy 60-second clips leave out the messy middle, the privilege, and the real hustle behind the scenes. So, before you start questioning why you're not one of "them," or why you can't just do what they do, hold off on feeling down, because mindlessly believing them will only pull you further away from the real truth.

Amid those influencers flexing posts, swirling trends, and suffocating peer pressure—all while juggling the constant craving for instant gratification—managing money feels like a real struggle. And to make it worse, you're pretty much left to figure it out on your own. Schools don't bother teaching financial literacy, and parents often seem too busy to guide you step by step—or, worse, they shut you out of money talks entirely, as if finances are some top-secret adults-only club.

So, here you are, trying to piece things together block by block, only to hit setbacks that could cost you a fortune. With this limited guidance, poor planning, and let's be real, some shaky self-control, it's easy to fall into the trap of overspending, struggling to tell the difference between what's necessary and what's just a fleeting want. And, as if budgeting isn't already a headache, society keeps shouting at you to invest and build wealth when, let's be honest, just covering the basics sometimes feels nearly impossible.

Do these scenarios ring a bell? Do they leave you feeling anxious or uncertain about how well you'll manage your finances? No need to worry—that's exactly where financial literacy comes in to keep your money game strong.

As you go through this book, you'll build your financial know-how one chapter at a time. No, this book won't turn you into a millionaire overnight, but it *will* give you the knowledge to make informed decisions throughout your life, and that is more important in the long run. No more stressing over where you went wrong, struggling to stick to a budget, or feeling lost in the sea of savings and investment advice. Financial literacy gives you the power to take control of your money, tackle setbacks, and seize opportunities as they come. Because at the end of the day, you should be the one controlling your money—not the other way around. No matter how much money is in your bank account, or even if you're still scrambling off the whole job thing, building your financial literacy is the only necessary initial investment you need to make.

When it comes to managing money, the whole concept of financial literacy might sound daunting. But with this book, it doesn't have to be; we'll start from scratch, so you're never left wondering how to grasp the basics of money management. While navigating adulthood feels a bit like being hit by a runaway train of responsibilities, learning the basics early gives you a head start, helping you to navigate life with confidence instead of constantly stressing over money. With the right financial knowledge, you can focus on what truly matters.

In this book, you'll find all the essential tools to budget smarter, save faster, invest early, and avoid common money mistakes. If managing finances once felt like wandering through a confusing maze, consider this book your step-by-step roadmap to financial success. Through the SMARTMONEY Framework, we'll breakdown everything into 10 easy steps:

- **S**—Success Mindset
- **M**—Money Basics
- **A**—Accelerate Your Income
- **R**—Rule Your Budget
- **T**— Treasure Growth
- **M**—Make Smart Spending Choices
- **O**—Own Your Credit and Debt
- **N**—Net Worth Building
- **E**—E-Security
- **Y**—Your Future Plan

The SMARTMONEY Framework teaches you how to earn, save, and spend wisely, while also helping you build wealth for your future, no matter where you are financially. Plus, you'll uncover game-changing shortcuts like creating a realistic budget that actually works, unlocking the power of compound interest for long-term growth, and mastering credit to steer clear of debt traps. Unlike those financial guides you may have stumbled upon, the SMARTMONEY Framework focuses on practical, real-world applications, tailored to your unique struggles and experiences. Think of this book as your wiser, more compassionate friend—someone who genuinely wants the best for you as you navigate healthier, more mature ways of managing money. Not another know-it-all adult dictating what you should do. Because, let's be real—what teenagers and young adults need isn't another authoritative figure lecturing them. It's real-world guidance that actually gets them.

At the end of the day, no one wants to constantly follow what others think they should do—you want your own freedom while building the confidence to handle responsibilities and challenges along the way. As a mother of two, beyond my career in finance and education, I strongly believe that financial literacy isn't just for adults—it's something that should start in the teen years, no matter what your background might be. By learning financial responsibility early, you're laying a foundation for a more financially secure and empowered generation. No more being hushed at home or ignored in school curricula—financial literacy should be accessible to all teens and young adults navigating adulthood. Whether you're juggling summer jobs, saving up for your first big purchase, or just trying to figure out how to "adult" with your finances, you need a guide that cuts through the noise and makes money management simple, clear, and actually useful. So take your time—let's tackle it one chapter at a time and start building a healthier, more stable financial future today.

CHAPTER ONE

[S]—SUCCESS MINDSET

> *Whether you think you can, or you think you can't—you're right.*

— HENRY FORD

Think about those moments when you truly believe in yourself—you feel confident, and you actually succeed. Now, think about the times when doubt creeps in—you hesitate, second-guess yourself, and end up failing. More often than not, success is about how you perceive yourself and your abilities rather than simply what you can or can't do. Your mindset plays a much bigger role than you might realize. It's not just a passing thought—it shapes your actions, behaviors, and, ultimately, your reality. The way you feel influences how you act, and how you act determines the outcomes you experience. The same holds true for your money mindset and financial reality. So, if you want a better financial future, start with transforming your mindset before it reshapes your reality.

THE POWER OF YOUR MONEY MINDSET

When it comes to financial literacy, you won't learn what you need to know just from books, classes, or online courses. While these forms of explicit education provide essential knowledge and tools for making informed financial decisions, our attitudes and decision-making processes around money are shaped by much more than that. No matter how much effort we put into learning about finances elsewhere, the ingrained effects of implicit financial socialization—especially from parents and family—often hold more sway. Even from the earliest stages of childhood, before you even know what money is, you're constantly absorbing attitudes, emotional connections, and habits around money. And while explicit education might be logical and sensible, implicit socialization often carries stronger emotional experiences, making it harder to override. On top of that, the reinforcement of financial values in daily life—especially from parents and older family members, who are often seen as authority figures—adds another layer of emotional pull, making familiar financial patterns feel almost second nature. Thus, it is key to understand your current money mindset, how it got its roots, and the ingrained influence by your family.

Inheriting Your Mindset

Has anyone ever told you how you have your mother's eyes or how you look just like your dad? Just as our parents pass down physical traits, they also "inherit" similar ideas and attitudes toward money from their parents. Did your parents encourage you to save your allowance in a piggy bank? Did your sibling ever warn you not to make those in-game purchases? Or did your grandparents share stories about their financial struggles when they were your age? It all starts with family—the first space where we learn values, behaviors, and beliefs. Not only are we influenced by their advice, we also observe the way family members handle money— the conversations they have and the choices they make, whether intentional or not, in everyday life.

It's no wonder our money mindset often bears a strong resemblance— or even an exact copy—of our parents' teachings and attitudes toward money. This explains why people frequently find themselves repeating financial patterns similar to those their parents created (Klontz et al., 2010). For example, a boy who grows up in poverty might develop the belief that "Wealth isn't meant for people like me." Later in life, this mindset becomes ingrained, leading to underachievement and low motivation to change his financial path. He most likely ignores opportunities because he has already accepted defeat when it comes to building wealth. On the other hand, a boy raised in a wealthy family is more likely to internalize the belief that "I was born to be rich." This mindset can make him more resilient, viewing setbacks not as failures but as steps toward even greater wealth.

Beyond your immediate family, your peers at school or in our social circles can be the second most influential factor in shaping your money mindset (Taneja, 2014). Think about those times when your friends are talking about the latest shopping trends or gadgets, making you feel pressured to keep up. While parents lay the foundation for our financial habits, those values and beliefs can shift as we expand into wider social circles (Danes & Dunrud, 1993). So, whenever you find yourself struggling with money management, take a step back and consider the influence of your typical social circles—not just your family. However, it's never too late to transform the way you think about money. As you read this, you're already starting the process of determining which financial practices and beliefs you wish to keep, and which ones need to be challenged to have a more secure, growth-oriented financial future.

MONEY BELIEF TRANSFORMATION

Transforming your money beliefs doesn't happen overnight, as they are deeply ingrained in your mindset. It takes some serious self-reflection—like, deep diving into your own head—and getting educated to even see those limiting beliefs. Moreover, when the mindset is shaped by those misconceptions and myths about money that have been passed down through generations, from your grandparents to you, it might have some deep emotional connection that you've never realized. These misconceptions likely have replaced the true principles of financial success, holding you back from reaching your full potential. So, take time to reflect on your upbringing. What did your family teach you about money? Which money beliefs need to be challenged? How do you feel when you think about money? Writing these beliefs down makes it easier to debunk them and discover a better, more effective mindset to manage your money and unlock your full potential. Beyond unlearning and re-learning those outdated myths, you also need to figure out whether you've set yourself up for a scarcity or abundance mindset about money.

Debunking Money Myths

- **Money grows on trees.** If you still think of money as an endless resource, like something that grows on trees, this might be a good sign you haven't been exposed to enough open conversations about money yet. As kids, it's easy to believe this when you see adults effortlessly whipping out their wallets with a wad of cash inside for groceries, meals, and shopping. It seems like money just magically appears whenever it's needed. But, in reality, money is earned through hard work—not just for the time spent doing a job but also as compensation for skills and effort. Think about when your parents asked you to do chores in exchange for screen time or that item you were eyeballing at the mall. It shows that rewards in terms of money don't come for free.
- **You have to be rich to save money.** We're constantly told we need to save money. But, you might be thinking, *What on earth can I save when I barely have enough to cover my needs?* Then, you end up with the mindset that saving is only for the rich because they always have extra money to spare. Actually, you don't have to be rich to start saving. Remember that feeling of being overwhelmed by a tough exam in school? At first, it felt overwhelming, but once you started by answering the easiest questions, you gained momentum. Before you knew it, you were nailing the test. So, instead of getting overwhelmed by the idea of saving thousands of dollars, focus on saving even a tiny bit—a dollar or two—from your allowance, gift money, or whatever you earn from chores helping out, or earnings from your part-time job. Saving is for everyone, not just the wealthy.
- **Credit cards give you free money.** As kids, we often see adults grab whatever they want from a store, head to the cashier, pull out a piece of plastic called a credit card, and

they're good to go. Credit cards can make it seem like you can get anything you want without spending a dime of your own cash. But, sorry to burst your bubble—credit isn't free money you can use whenever and for whatever you want. A credit card is actually a payment tool that works like a loan, and, of course, that loan has to be paid back, often with sky-high interest. Unlike paying with cash, where the transaction is done and over with, using a credit card means you're borrowing money, and the longer you leave it unpaid, the more it grows with interest. Indeed, it's so easy to get caught up in the swipe-and-buy thing, but know that your wish list could end up costing you way more than you think—like, double or triple the item's price.

- **Money isn't a concern until you're grown up.** You probably think you don't have to worry about managing money since you're not working yet. That's just for adults with full-time jobs, right? Well, even if you don't yet have a steady paycheck, it's never too early to start learning how to manage money. As a young adult, you definitely don't want to find yourself gasping at the end of the week, just waiting for your next allowance or paycheck to roll in. Or worse, relying on a credit card every time your money runs out, only to end up drowning in debt. Learning about money management early helps you avoid those costly mistakes when you do become an adult and sets you up for success with smart financial habits. Think about the habits you already have—like checking your phone first thing in the morning, exercising every other day, or grabbing a coffee before school or work. Just like those, good financial habits can become second nature, something you do without even thinking about it.

- **All debt is bad.** When you begin working and earning your own money, you'll likely hear that debt should be avoided at all costs, as if it's the most terrible thing you could ever have. Indeed, debt used for unnecessary, discretionary spending is harmful and should be avoided. But, in some cases, debt can also be necessary for achieving significant future goals. For example, student loans can finance higher education, and a mortgage can enable homeownership, as long as you can manage repaying your debt in a reasonable amount of time. You should make mismanaging debt the enemy, but not the debt itself. In the future, if you ever need to borrow money, the most important thing is to use it wisely, only for things or experiences that serve a fulfilling and long-term purpose. Besides, you must fully understand all the terms tied to the debt and its repayment to ensure it doesn't put unnecessary strain on your finances. We'll talk about this more in Chapter 7.

Scarcity vs. Abundance Mindset

While debunking the money myths focuses on each specific misconception, transforming your mindset focuses on shifting the entire belief system you've held about money toward a more optimistic and growth-oriented perspective. Becoming aware of your money mindset helps you understand why you made the decisions you've made or why you think the way you do. Without even realizing it, our money decisions have often reflected the mindset we've developed over time. So, it is just a matter of time before you can identify yourself as leaning more toward scarcity or abundance. Once you've identified your mindset, you'll better understand your financial patterns and how to reshape them for a more secure future. Below are the characteristics of each mindset to help you figure out which one you are leaning more toward:

- **A scarcity mindset:** This is the belief that your financial, emotional, and social resources are limited. With this mindset, you most likely fixate on things you lack, making you feel as though you have no possible opportunities to get what you want. It's no surprise, then, that this mindset often leads to worrying about having enough money, resisting the idea of investing, or even feeling guilty about spending money, even on basic necessities. And when you do have money, a scarcity mindset convinces you to be anxious about whether you'll earn more, reinforcing the belief that financial success is always out of reach. It can also create envy and resentment toward others who seem to have more financial stability. If you frequently find yourself thinking, *I'll never be rich, I can't afford that, I'm always falling behind,* or *What if I lose everything?,* chances are you're operating with a scarcity mindset.

- **An abundance mindset:** This mindset focuses on the ample resources and opportunities available throughout life. It creates constant optimism and confidence to build a stable financial well-being. Beyond exuding confidence, having an abundance mindset shows a willingness to invest and explore possibilities in new opportunities. People with an abundance mindset are more likely to seek partnerships and be proactive in growing their wealth. With the right approach of seeking possibilities and being willing to explore, they believe that financial stability and security are within reach. Instead of excessively worrying about having enough money, people with an abundance mindset take control of their finances and perceive money as a tool that can be managed and multiplied rather than as a limited resource. As a result, they are more than willing to explore ways of investing and wealth-building. Those with an abundance mindset are likely to think, *There are always opportunities, I am grateful for what I have, Money is a tool to create the future I want,* or *I am willing to take calculated risks for greater reward potential.*

If an abundance mindset resonates more with you, then you are well on your way to allowing yourself to explore new, unfamiliar possibilities for greater reward potential. Still, you must make sure you have carefully calculated and balanced the potential rewards and losses whenever trying new things, whether it's choosing a new savings plan, investing, or pursuing a new job. Focusing on available opportunities does not necessarily mean taking advantage of all of them blindly without giving a second thought to how they might turn out.

But if the scarcity mindset feels more like your usual approach, then it's time for some changes in how you think about money and manage it. A scarcity mindset often does more harm than good and can hold you back from attaining your goals. Cultivating an abundance mindset starts by focusing on what you have, not just what you lack. Instead of dwelling on not having a job, you're better off using that time to explore new skills and knowledge you've been hesitant to approach. And when money is tight, try thinking, *It's time to explore how I can earn or save more.* Most importantly, you need to start introducing positive affirmations such as, *I'm grateful for the money I have, I'm willing to learn how to earn and manage money wisely,* and *I'm destined for financial success.*

If you've ever wondered how or why a friend approaches money with positivity and optimism, pay attention to their mindset. And, try to surround yourself with positive like-minded people when you can, so you can adopt that abundance mindset. Hanging out with people who share a scarcity mindset won't help you to develop a healthy relationship with money, so maybe it's time to join a new circle and find your new tribe.

FINANCIAL CONFIDENCE BUILDING

The transition to managing personal finances is a leap that most of us—especially teens and young adults—struggle to start. You might think, *Wait, is no one handing me an allowance anymore? So I have to figure out how to earn money and manage it all at once? Am I really expected to make financial decisions, not just for now, but for the future, too?* For many, the idea of budgeting, saving, or investing feels unfamiliar, even intimidating. So, what can make this seemingly harsh transition feel a little smoother and less scary? Building your confidence in managing money through experience and perseverance. It's essential, even before you start earning your own money. Because, let's be real—financial confidence isn't an innate trait; it's a trained skill.

Gradual Progress

The key to developing good money habits doesn't include magically saving hundreds of dollars in a few months, impulsively checking off every item on your wish list, or making high-stakes stock market investments within your first few months of working. If these overnight successes come to mind while thinking of personal finances, you need to shift your perspective and steer clear of those influencer posts promising instant wealth. These exaggerated expectations will only leave you feeling discouraged and stuck in a cycle of "never enough," pushing you away from engaging with your finances altogether. Instead, know that the journey to financial literacy is a gradual process, one filled with small wins, occasional setbacks, and continuous learning.

While financial literacy is a journey, setting goals for where you want your finances to be should be the most important step you take. It's like figuring out each base camp before you climb a mountain or learning the basics of cooking before trying to prepare a gourmet meal. Setting financial goals helps build confidence because you'll clearly see the small steps and wins along the way before reaching your bigger, loftier goals. Instead of vague anxieties, having financial goals gives you a clear target to aim for, followed by specific actions to achieve each one. Without goals, your journey to financial stability is merely wishful thinking. Besides setting goals, breaking larger ones into smaller, more attainable steps allows you to see tangible progress, providing positive reinforcement along the way.

When first talking about financial goals, do you ever catch yourself thinking things like, *I want to be rich, I want to buy whatever I want, I want to save more money*, or *I want to reach financial freedom*? Stop with these vague goals because they won't get you far. Instead, let's get real about what you really want and break it down into smaller, doable steps. Instead of simply planning to "save more," think about specifics like "cutting back on spending money on hobbies next month" or "putting $20 into savings each paycheck." While being specific, you must also keep goals aligned with your current financial situation. Unrealistic goals only bum you out in the long run and make you feel like you're never gonna get there. Once you've got your goals sorted, it's time to put a timeframe on them. No more wishy-washy "someday" or "in the future" vibes. Be specific.

The Hazards of Imposter Syndrome

As you check off your financial goals, it's easy to fall into the trap of feeling like you don't deserve your success. Maybe you finally saved up for a new gaming console, scored tickets to a concert, or started making money from your side hustle—whether it's babysitting, dog walking, or selling crafts. If, after hitting your financial goals, you find yourself downplaying your achievement, thinking, *Anyone could have done that*, or *It was just luck*, then the imposter syndrome might be kicking in. If you're not familiar with the term, *impostor syndrome* is a feeling you have when you doubt your success, causing you to constantly fear being exposed as a fraud. It makes you believe your competence is due to luck, timing, or other external factors, rather than your skills or efforts. This feeling can get worse when you compare yourself with others or avoid asking questions because you're afraid of looking incompetent.

When these thoughts hit hard, remember: You're not alone. Everyone, no matter their age, has felt this way at some point. To combat it, keep a record of your achievements and the effort it took to reach them. These records help to remind you that you made those things happen. If certain posts or comments online fuel your self-doubt, you should distance yourself from them and replace those thoughts with positive affirmations. Most importantly, you must understand that it's okay not to know everything—learning is a process, and you're exactly where you need to be.

Interactive Element

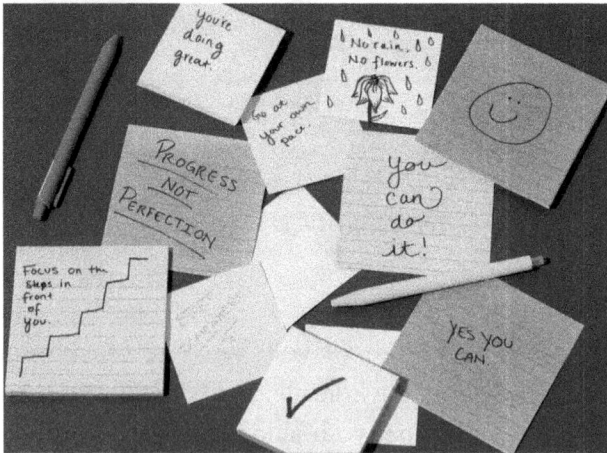

Transforming how you think about money hinges on how well you can reflect on your thoughts and emotions about money. To truly understand your thoughts and feelings about money, take a moment of quiet reflection and write down any negative thoughts you have about money. Maybe you catch yourself thinking, *I'm just bad with numbers, I'll never be able to afford nice things like other people can*, or *My friends and family always pressure me to buy things I can't actually afford.*

Once you've listed your thoughts out, challenge each one by replacing it with a more empowering statement, such as *I can always learn to manage money, It's unfair to compare my financial journey to others—we all have different starting points and goals*, or *I have the power to make my own financial choices*. When you're done, write each motivating statement on a sticky note and place them somewhere visible—your mirror, your desk, or on your phone as wallpaper—so these new, empowering beliefs become second nature. The more you reinforce them, the stronger they'll shape your financial mindset.

WHAT'S NEXT

Now that you've shifted your mindset and started perceiving money as something you can control, the next step is to master the basics of personal finance. In the next chapter, you'll dive into the nature of money, inflation, and banking—essentials for kickstarting your financial journey. It all begins with understanding these fundamentals to know why and how to build wealth over time to keep up with inflation, starting with a solid grasp of your bank accounts.

CHAPTER TWO
M—MONEY BASICS

You've got to tell your money what to do, or it will leave.

— DAVE RAMSEY

Almost 100% of millennials use a mobile banking app for daily purchases (White, 2020). But does simply using an app mean they truly understand how to manage their finances? Many check their balance and transfer funds—and nothing else. Many have no idea about hidden fees, missed opportunities to grow their money, or whether their account is even secure. Some don't even realize they're at risk of fraud until it's too late. Thus, in this chapter, we'll cover the fundamentals of banking and how to make the most of your accounts, helping you to build a strong and secure financial foundation. No more relying on your parents to handle transactions or learning through costly trial and error.

WHAT IS MONEY?

For all of us, money is likely one of the first things we can't leave the house without. Whether it's buying a coffee, purchasing groceries, or renting your first apartment, money has become a fundamental necessity in life. But what exactly is money? It's not just the cash in your wallet. Money is a medium of exchange that society collectively agrees upon to make transactions easier. If you think financial transactions only revolve around cash or credit cards, keep in mind that the history and evolution of money stretches far beyond that.

Bartering

Before money even existed, people conducted transactions by exchanging goods to get what they needed, a system known as *bartering*. Bartering meant swapping what you had for what someone else had, but only if you both wanted each other's stuff. Consider a potato farmer that is looking to make a trade for fruit. If the fruit farmer also wants potatoes, then they've got a deal. While bartering provided a way for people to exchange goods, it had a major limitation—it required a "double coincidence of wants." In other words, both parties had to want what the other was offering. For example, if the fruit farmer wanted potatoes, but the potato farmer didn't need fruit, then the trade couldn't happen.

Commodities and Precious Metals as Money

To address the inefficiencies of bartering, people began using commodity money—items that held intrinsic value—ranging from ax heads, knives, and cowrie shells to salt and whale teeth. Unlike bartering, commodity money could be traded for other goods, even if the recipient didn't directly need what the other person had. Besides, it also made long-distance trade easier. However, while commodity money introduced more convenience and a more uniform way of trading, it still had limitations. Many of these items deteriorated over time and were difficult to transport, especially in large quantities.

To overcome issues related to commodity money, societies turned to precious metals such as gold and silver as a more stable form of money. Unlike perishable commodities, precious metals didn't degrade over time. While commodities varied in size and quality, precious metals offered a consistent value because their weight could be verified, making trading more efficient and fair. Initially, precious metals were used as currency in the form of bullion or bars. However, to simplify exchanges, bullion gradually transitioned into coinage. The Kingdom of Lydia was among the first to produce coins, minting electrum pieces around the year 630 (Holt, 2021). Around the same time, in 640, China also began producing spade-shaped bronze coins, with a minting site discovered in 2021 (Zhao et al., 2021).

As economies evolved, so did the need for an even more efficient form of money, leading to the use of paper money or banknotes. Transporting large amounts of precious metals was cumbersome due to their weight, and carrying them also attracted thieves. To solve this problem, people began depositing their precious metals with trusted institutions, which issued paper receipts that could be used for trading, laying the foundation for the modern banking system.

The Rise of Paper Money

Around 1260, China's Yuan dynasty started to make the transition from using precious metal coins to paper currency. While deposits of precious metals mostly backed early forms of paper money, the Yuan dynasty gradually shifted toward a fiat system, where currency production was based on government decree. It wasn't until after 1271, when Marco Polo visited China and became fascinated by the sophisticated monetary system the country had built, that paper money started to take off (Vogel, 2013). China was already a step ahead of the rest of the world, ensuring the emperor controlled the money supply and its various denominations while mandating its acceptance for all transactions. Centuries later, the concept of state-backed paper currency, rather than metal-based money, spread to Europe, laying the groundwork for the banknote system we use today. By this time, people had realized that money didn't necessarily need intrinsic value as long as it was widely accepted for its purchasing power. Think about what makes money different from an ordinary piece of paper—it's all about a collective agreement.

Credit Cards and Modern Currency

Until the 20th century, paper money remained the dominant form of currency. However, with the rise of digital transactions toward the end of the century, credit cards emerged as a major financial tool. Credit cards offer a more convenient alternative than cash, which can be bulky and risky to carry in large amounts. Beyond simply enabling direct exchanges like paper money, credit cards introduced the concept of borrowing money from banks. Also, with the growth of the internet and smartphones, financial transactions began shifting from physical to digital platforms. Mobile banking and digital payment systems like PayPal, Apple Pay, and Venmo made it possible to complete transactions without needing a physical card—just a phone.

At the start of the 21st century, transaction methods continued to evolve rapidly, and cryptocurrencies emerged as a potential replacement for the traditional fiat system controlled by governments and central banks. Cryptocurrencies operate on decentralized blockchain technology, making them more secure, transparent, and borderless, eliminating the need for banks or financial institutions. Have you ever heard of Bitcoin, Ethereum, XRP, or Dogecoin? These are all examples of cryptocurrencies. With the rapid rise of contactless payments, the need for physical cash is steadily decreasing. It's no wonder we might soon see a completely cashless society. Understanding the history of money helps us to see how quickly financial systems evolve—especially in recent centuries—making it crucial to keep up with trends and thoroughly understand them before diving in.

THE CONCEPT OF INFLATION

Do you remember the first time you ever watched a movie in a theater? Now, think about how much a movie ticket cost back then compared with today. You don't even have to go back decades—just look at the past 10 years. In 2014, the average movie ticket in the US was $8.17. Fast forward to 2024, and that price has jumped to $11.31 (The Numbers, 2023). A difference of about $3 might not seem like a big deal at first, but that's actually a 38% increase in just one decade. Crazy, right? Now, think beyond movie tickets—this same price hike is happening everywhere. Concert tickets, takeout, groceries, rent, even real estate—they're all creeping up, making your hard-earned money worth less over time. Well, this is how inflation acts by slowly chipping away at your purchasing power, little by little, year after year.

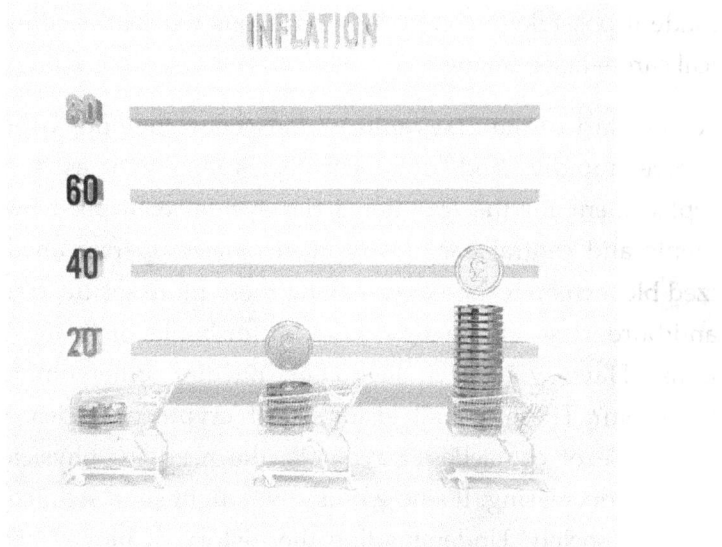

Today, inflation might not seem like a big deal, like it's something far off in the future. You might even wonder why we're talking about it while learning the basics of finance. But here's the thing: Understanding inflation now prepares you for what you'll definitely face in the near future. It's not some distant, lurking threat; it's a real challenge that affects your money every day. The sooner you get familiar with it, the better you can manage your finances and avoid getting caught off guard.

So, what exactly is inflation? Imagine blowing up a balloon—it gets bigger and bigger until, suddenly, pop! Inflation is similar—it's the steady increase in the price of goods and services over time. But, unlike a balloon that might deflate, inflation only goes one way: up. Your money feels like it's shrinking, and that's not a good thing. Even worse, while prices keep climbing, your paycheck doesn't always grow at the same rate as inflation. While the cost of everything around you goes up—groceries, rent, gas, even your favorite fast food—your income might stay the same. This makes it harder to afford the same things you used to, and for many people, it leads to borrowing money just to keep up. It's like running on a treadmill that keeps accelerating. You're working hard, but you're not getting any closer to your goal. And if you think you're safe by just saving money in a bank account and not spending it? Think again. Inflation still eats away at your savings. The $100 that once filled a grocery cart when you were a kid now barely covers a few bags of essentials.

Supply and Demand

If inflation makes life harder for everyone, why does it still happen? Well, inflation isn't always bad—a steady, controlled level of inflation is actually necessary for a healthy economy to grow. Inflation happens when the demand for products—whether goods or services—outpaces the supply. Think of five hungry boys who all want a burger, but the shop only has enough ingredients to make three. Since there aren't enough burgers for everyone, the shop decides to raise the price, knowing that some of the boys would be willing to pay more. When there's a high demand but not enough supply, prices naturally go up, causing inflation (Chen, 2022). This same concept applies to everything—from gas and groceries to sneakers and concert tickets. If supply can't keep up with the demand, businesses increase prices, and that's how inflation keeps rising.

While inflation happens for a variety of reasons, from economic shifts to government policies, the cause of demand outpacing supply is one of the most significant yet inevitable factors. With the world's population constantly growing, the need for goods and services will never slow down. So, if you think inflation isn't something to worry about, think again. The longer you wait, the more your money loses its purchasing power, struggling against rising prices and skyrocketing interest rates. There's no such thing as starting too early when it comes to managing your finances. No matter how young you are or how little you have in your bank account, the best time to take control of your money is right now.

BANKING ESSENTIALS

As you begin your financial journey, opening your own bank account becomes a necessity. Think of it as an upgrade from the piggy bank sitting on your desk—except this one keeps your money safer and helps you manage it better. Unlike a piggy bank, which you can easily crack open for a quick ice cream run, a bank account reduces the temptation to spend impulsively. With your money stored safely, you're more likely to pause before splurging on impulse purchases or unnecessary shopping.

Having your own bank account also comes with a new level of responsibility. You'll quickly get used to checking your balance, understanding withdrawals, and tracking where your money actually goes. More importantly, it encourages you to build good saving habits, whether you're working hard to buy a new phone, pay for your future college tuition, or even moving out on your own someday. And when you start working consistently, most employers require a bank account for direct deposits. Even if you don't have much money now, opening a bank account sets you up for success and helps you to become more financially prepared for the real world.

Choosing Your First Bank

When choosing a bank to open your first account, start by asking your parents if they've already set up a child or teen account in your name. Some parents open accounts early to deposit gift money or savings for you. If they haven't, it's time to shop around for banks that offer special accounts for teens before transitioning to a standard adult account. Having an account tailored toward your age comes with key benefits, such as lower fees, financial education resources, and parental oversight. The details of specific protections and features might vary among different banks, so make sure you research and compare each one.

To start, check out local banks in your community that offer customer-friendly options and services designed for young account holders. No matter which bank you choose, make sure it's FDIC-insured, ensuring that your money is safe and protected. Aside from security and account features, consider accessibility based on your needs. If you prefer handling transactions in person, look for a bank with a physical branch nearby. But if you'd rather manage everything from your phone, ensure the bank has a reliable mobile app and user-friendly online banking features. Also, check whether the bank has fee-free ATMs within your area to avoid unnecessary withdrawal charges.

Checking vs. Savings Account

When opening your first bank account, you will need to decide between a checking or savings account. Checking accounts are designed for everyday transactions, providing easy access to your money through debit cards, checks, and ATMs. They are ideal for daily use but typically offer little to no interest, making them less beneficial for long-term savings. On the other hand, a savings account, as the name implies, is meant to store your money while earning interest over time. Unlike checking accounts, savings accounts likely come with withdrawal limits, and exceeding them can result in fees. While a checking account is best for regular trans-actions, a savings account is essential for setting aside money for large purchases or emergencies.

Each type of account has its own advantages and limitations. Ask yourself: *What do I need this account for? Which option best suits my current situation?* Starting with a checking account may be more practical, especially if you expect to receive a regular paycheck into your account. However, be sure to use this account wisely and avoid impulse spending. Once you accumulate more money than you need for daily expenses, consider opening a savings account to encourage financial discipline and limit unnecessary withdrawals. Ultimately, having both a checking and savings account allows you to manage your money effectively while also growing your savings.

Whether you decide to open a checking or savings account, you must prepare some paperwork for your bank application. While requirements may vary by bank, you will typically need a form of identification (a state ID card, driver's license, or passport), your Social Security Number, proof of address, and an initial deposit (if required). If you're under 18, you may also need a school ID (if available) as well as a parent's or guardian's signature and ID.

Avoiding ATM Fees

Have you ever withdrawn money from an ATM, only to notice that your balance decreased by a few extra dollars on top of your withdrawal amount? That's likely because you used an out-of-network ATM that doesn't belong to or is not partnered with your bank. When withdrawing from an out-of-network ATM, you're typically charged two fees—a surcharge from the ATM provider (averaging around $3.15) and an out-of-network fee from your bank (averaging around $1.58) (Bennett & Goldberg, 2019). These fees might not seem like much at first, but if you withdraw money multiple times a week, they can add up to hundreds of dollars per year—money you could be saving instead.

While using an out-of-network ATM might be unavoidable in some situations, you shouldn't make it a habit. So, when choosing a bank, consider the availability of fee-free ATMs in your area. If you already have a bank account, check your bank's website or mobile app for an ATM locator—simply enter your ZIP code to find the nearest fee-free ATM. If no in-network ATMs are nearby, having a withdrawal plan to take out your money might be efficient—for example, instead of withdrawing small amounts multiple times, take out a larger sum in a single transaction to minimize the fees. But keep in mind to check your bank's policy first to ensure you don't exceed the limit, if there is one, when withdrawing a larger amount.

Learning Mobile Banking Basics

As we all rely on quick and convenient transactions for everyday purchases, mobile banking has skyrocketed in popularity, especially with banks offering user-friendly interfaces. With just a few taps on your phone, you can send money, pay bills, and shop online anytime, anywhere. But mobile banking isn't just about making transactions—it's a powerhouse tool for managing your money. You can check your balance, transfer funds, review your transaction history, track your spending habits, and even set up parental controls to monitor real-time spending. As long as you have a stable internet connection, your banking information is always at your fingertips. Beyond convenience, mobile banking also gives you faster access to customer support, helping you resolve issues without making a trip to the bank. And let's not forget security—most banking apps are protected with multiple layers of security, like passcodes, biometric authentication (fingerprint or face recognition), and one-time passwords (OTPs) to ensure your money stays safe.

However, while mobile banking is super convenient, it does come with some risks, especially for those who aren't tech-savvy and those who are first-time users and therefore could be more vulnerable to online scams. First, take the most important precautions by creating a strong, unique password. Avoid using easy-to-guess information like your name, birthdate, simple number sequences, or your school name. And whatever you do, never share your password with anyone. If you receive emails, texts, or phone calls asking for personal or financial details, you must stay cautious. Keep in mind that legitimate banks would never ask for your password or authentication codes. Also, be mindful of where you log in. Public Wi-Fi networks often lack strong security measures, making them more vulnerable to hackers, so it's better to stick with your mobile data or a secure private Wi-Fi connection. Most importantly, make it a habit to check your account balance and transaction history regularly. The faster you spot suspicious activity, the quicker you can take action to protect your money.

BANKING ACCOUNT OPTIMIZATION

Unlike the piggy banks you've been using for years, which simply store your money, having a bank account opens doors you might not even know exist. Having a bank account allows your money to earn interest and provides numerous other benefits. While fees and accessibility are important factors, choosing a bank that offers additional perks can make a big difference. Understanding these benefits helps you to select a bank that not only keeps your money secure but also works in your favor. There are a few bank benefits you might want to consider taking advantage of.

Finding Hidden Bank Perks

There's more to opening a bank account than getting a debit card and using it for purchases. There are plenty of hidden perks within your account that you might not even be aware of. By taking advantage of these benefits, you can make your money work smarter for you. Here are some valuable perks worth exploring:

- A **high-interest savings account** allows your money to grow much faster, sometimes earning up to a few times more in interest compared to a traditional savings account. With banks competing for new customers, many offer attractive interest rates to win you over. If you're considering opening a new account, you should start looking into online banks in particular—they can offer higher interest rates because they save on costs by not operating physical branches. To find the best option, compare interest rates from different banks. Be sure to check whether a high interest rate is a temporary promotional offer or a long-term benefit.

- If you frequently use your debit card, opening a **rewards checking account** could earn you cash back on everyday purchases. In addition to getting cash back on purchases, these accounts often come with perks like higher interest rates and ATM fee reimbursements—benefits that standard checking accounts typically don't offer. However, to qualify for these rewards, you'll likely need to meet certain requirements, such as keeping a minimum balance, completing a set number of transactions per month, setting up direct deposits, or signing up for online banking. If these conditions align with your current spending habits, a rewards checking account could be a great way to maximize your money. Otherwise, you might be better off sticking with a standard savings account.

- Many banks offer introductory bonuses as an incentive for opening a new account. These bonuses often require you to set up direct deposits or maintain a minimum balance for a specific amount of time in return for a cash reward. Both traditional and online banks have different terms and conditions, so it's better to shop around first to find the best deal. Before opening an account, be sure to check for any hidden fees that could reduce the amount of your bonus.

While the upfront reward may seem attractive, consider the long-term benefits as well. Most importantly, make sure you meet all the bonus requirements within the specified timeframe to receive the full reward.

- If you're torn between opening a checking or savings account, a **money market account (MMA)** might be the perfect middle ground. MMAs offer the best of both worlds —higher interest rates than regular savings accounts, plus the flexibility to access your money like a checking account. Many MMAs even come with debit cards, making everyday transactions more convenient. However, MMA interest rates can fluctuate due to money market conditions, and these accounts usually limit the number of withdrawals or transfers you're allowed to make each month. Exceeding this limit may result in fees, so be sure to stay within the allowed transactions to avoid extra charges. If you plan on using your debit card for everyday purposes, an MMA may not be the best account for you.

Setting Automation Features

Trying to simultaneously manage your bills, cover basic expenses, and save money can feel overwhelming. Fortunately, automation features in your bank account can help simplify things. With auto-mated payments, you don't have to manually pay each of the bills yourself, reducing the risk of late fees while giving you the peace of mind that your expenses are always paid on time. Just be sure to schedule payments before their due dates to avoid any penalties or fees. Automation can also help build your savings effortlessly by automatically transferring a portion of your income into your savings account. It ensures you never forget to set money aside or get tempted to spend it on non-essential purchases. Once you set up automation, you're already taking a step toward better financial discipline by making your bill payments on time and consistently saving. The process for setting up automation can vary by bank, so you should check with your bank directly or explore its mobile app to see what options are available.

Linking Bank Accounts

If you already have both a checking and a savings account and frequently transfer money between them, linking your accounts can make transactions faster and more convenient. Linking bank accounts means connecting multiple accounts, allowing you to transfer funds easily, split direct deposits, and automate your savings. One major benefit of linking your accounts is avoiding overdraft fees. If your checking account balance is at risk of dropping to below zero, funds can be automatically transferred from your linked savings account to cover the shortage.

If both accounts are with the same bank, the linking process is usually simple and can be set up through the bank's website or mobile app. Some banks waive monthly maintenance fees if you have linked accounts and maintain a minimum balance across them. If your checking and savings accounts are at different banks, you can still link them, but you'll need to provide your account and routing numbers for verification. Once linked, transfers between banks might take a few extra days to process. To keep your linked accounts secure, always use a strong password and enable multi-factor authentication.

THE DIGITAL MONEY ECOSYSTEM

You've installed one of these apps, like Cash App, Apple Pay, PayPal, or Venmo, right? Among all generations, you, as Gen Z, are widely considered the generation that's not just tech-savvy but is defining tech. You're the most educated and diverse generation yet, and companies across every industry are scrambling to keep up

(Parker et al., 2019). That's why digital wallets have exploded—they're not just convenient; they're built to cater to your need to make sending and receiving money faster, easier, and less stressful than ever.

And let's be real: You're way more comfortable shopping online than previous generations, especially when it comes to those discretionary purchases. It's no surprise that over 40% of Gen Z and a whopping 50% of Millennials are planning their discretionary spending week by week (Costa & Gardener, 2021). Especially with the rise of the "buy now, pay later" feature, digital wallets aren't only a convenience—they're changing the game. By now, you probably find yourself relying on them, transforming how you transact with merchants. However, simply using them isn't enough, because although these digital wallets come with their super convenience, they also come with greater risk. Therefore, you should know how to leverage them strategically to improve your spending patterns while staying safe from potential risks.

The Benefits of the Digital Wallet

While mobile banking is essentially an extension of your bank account, digital wallets focus mainly on facilitating payments. They store or link to your payment information, allowing you to make purchases without needing your physical debit or credit card. And, unlike traditional banking apps, digital wallets are often accessible not just from your phone but also from a computer or tablet. With a digital wallet, there's no need to carry a stack of cards everywhere you go. Your payment details are stored securely, so you can tap, scan, or send money in seconds. Plus, many digital wallets can store more than just debit and credit cards—they can also hold gift cards, membership cards, coupons, and loyalty rewards. That means not only added convenience but also potential savings every time you shop!

Because digital wallets depend on your phone to complete transactions, always make sure you have cellular data enabled when heading out, as Wi-Fi isn't always available. Using your mobile data is also a safer option—public Wi-Fi networks can be risky, making it easier for hackers to access your information. As your banking details are stored in the app, losing your phone or getting hacked could put your money at risk. Therefore, to make sure you stay protected, what matters most is creating a strong, unique password and enabling two-factor authentication (2FA) for both your device and the digital wallet app. Then, you should set up spending limits to prevent overspending, which can happen easily with the convenience of digital payments.

Always Carry Some Cash

Although you likely rely on a digital wallet for most of your payments, it's still a good idea to carry some cash. Some stores or certain purchases do not accept digital payments, especially for very small purchases, so having a backup ensures you're never stuck without a way to pay. Trust me—there are still those random moments when only cold, hard cash will do.

Cryptocurrency

Beyond digital wallets, cryptocurrency has emerged as a new way to make transactions, especially among younger people like you. While its skyrocketing, and sometimes crashing, value often grabs headlines, the real idea behind cryptocurrencies came from a simple problem, making global transactions easier. Back when the internet was still new, the idea of digital money seemed strange. People were used to cash and physical bank transactions. But as technology advanced and more of our lives moved online, people became more comfortable with cashless payments, opening the door to digital currencies—and, eventually, cryptocurrencies.

The Benefits of Cryptocurrency

Now, you might hear terms like *digital currency* and *cryptocurrency* and assume they're the same, but they're actually two totally different things. Digital currencies are any form of money that exists electronically, like the money in your bank account or digital wallet. They're still tied to real-world currencies (like USD or Euros) and are regulated by banks or governments. Meanwhile, cryptocurrencies aren't connected to any country or controlled by any government or financial institution. They exist only online and don't have physical coins or bills. Instead, they're secured by cryptography, which encrypts and verifies transactions to prevent fraud. Unlike digital currencies, cryptocurrencies run on a decentralized network called *blockchain*. Think of it as a super-secure, transparent, and unchangeable online ledger where every transaction is recorded. This system eliminates the need for banks and makes transactions more direct—but also more volatile and risky.

You've probably stumbled across headlines about people becoming overnight millionaires thanks to the skyrocketing prices of cryptocurrencies—sometimes from coins they didn't even realize they owned. Bitcoin, Ethereum, Litecoin, and Ripple are some of the most well-known cryptocurrencies, and despite their ups and downs, they've managed to hold value over time. If you are ever interested in exploring cryptocurrencies, you might want to visit some centralized exchanges as intermediaries to trade between buyers and sellers. Some of the most well-known platforms for buying cryptocurrencies are Coinbase, Binance, and Kraken. There, you'll find a wide range of cryptocurrencies, their trading activities, and features. After creating an account on these sites, you might be required to provide personal information and upload identification documents. Once your account is funded through your chosen payment method, then you can place an order to buy cryptocurrency.

Cryptocurrency is stored entirely in digital form, and once you have some, you might be tempted to make transactions for online purchases. Those in tech or digital services might lean more toward accepting cryptocurrencies for seamless global transactions. Other than goods or services related to tech, you might be able to buy games, music, movies, and other digital content with cryptocurrency. However, paying through cryptocurrencies can depend entirely on the specific cryptocurrency and your location.

The Dangers of Cryptocurrency

Although cryptocurrencies have the potential to be a game changer due to their fast and low-cost international transactions and their extreme increases over short periods, cryptocurrencies also come with significant potential losses. As they are entirely stored in digital form, cryptocurrencies are more vulnerable to hacking, scams, and theft, especially for beginners. If you lose access to your private key, you could lose your crypto permanently—there's no bank to recover it for you. Moreover, as cryptocurrencies are not controlled by specific governments or financial institutions, their rules and details evolve quite rapidly, making them more unpredictable than traditional currencies.

If you're thinking about investing in crypto, learning about specific cryptocurrencies and their market history is a must. Most importantly, keep up with the latest news and developments in the cryptocurrency world. Once you decide to own cryptocurrency, only invest money you can comfortably afford to lose, especially if you feel like your calculations are highly speculative. Then, you should stick to well-established and reputable exchanges with strong security measures. If you ever come across those get-rich-quick schemes with crypto, being skeptical is always a good idea. Don't let yourself get fooled by the fear of missing out (FOMO), which only leads to impulsive decisions.

WHAT'S NEXT

Now that you've built a solid financial foundation and learned how to set up and manage a bank account, it's time to focus on the next step—growing your money. Earning a paycheck from a full-time job is just one way to make money, but there are many other opportunities to increase your income and even multiply it. In the next chapter, we'll explore strategies to help you earn more, maximize your savings, and reach your financial goals even faster.

CHAPTER THREE

$\boxed{\text{A}}$—ACCELERATE YOUR INCOME

> *The quickest way to double your money is to fold it and put it back in your pocket... or start earning more.*
>
> — WILL ROGERS

The fear of not having enough money to cover your current needs—let alone future ones—isn't solved by simply avoiding purchases. Trust me—stuffing your cash back into your pocket and only pulling it out for absolute necessities won't get you very far. What truly makes the biggest impact is increasing your income—after all, your earning potential is limitless. No matter what skills you have—whether it's babysitting, mowing lawns, or creating something online that people are willing to buy—you already have what it takes to turn your time and effort into cash. So, let's dive into practical ways you can start earning today.

GENERATING INCOME

When people hear the word *income*, most think of it as the money they receive from working, but it's actually more than simply clocking in and out on time at a job. Income is what you earn in exchange for not just your time but also your skills, effort, and the value you bring to the table. If you're putting in your time, energy, and skills, it's only natural to expect fair compensation. Think of income as the lifeblood of your finances—maintaining a steady flow is essential for budgeting, saving, and investing. Now, you might wonder whether your current income truly reflects your worth. Essentially, your income is a direct reflection of the value you provide, so don't settle for less than what you deserve.

Having a job and earning money is about more than simply covering essential needs, saving for the future, or reaching financial goals. Especially for teens and young adults who are still fully supported by their parents, working isn't just a way to fund a wish list, go to concerts, or buy the latest video games—it's an introduction to responsibility and independence. Think back to your first paying job. Was it babysitting, shoveling snow, or mowing lawns in your neighborhood? How did it feel to earn money for your efforts? Even if it wasn't a consistent job, that early experience of working for pay planted the seeds of a good work ethic and self-development. Research shows that having a job at a young age helps to develop essential skills like independence, responsibility, and interpersonal communication, qualities that continue to grow into adulthood (Phillips & Sandstrom, 1990).

Developing a Good Work Ethic

A good work ethic isn't something you magically develop as an adult. It starts long before you land your first full-time job. Think about the times your parents asked you to do household chores, clean your room, or help with yard work in exchange for a little extra cash. That was likely your first lesson in trading time, effort, and energy for money, a concept that becomes even more relevant once you enter the workforce. One study found that teens who start working consistently—whether part-time or full-time—are more likely to complete a four-year degree compared to those with little or no work history (Staff & Mortimer, 2007). But even if college isn't in your plans, early work experience can help you discover a potential career path you truly enjoy.

Beyond developing a good work ethic and exploring potential careers, having a job teaches you just how much time and effort it takes to earn a paycheck. Understanding the struggles of making money can give you a new appreciation for what your parents have gone through to provide for your needs—and, eventually, what you'll experience as an adult. Even if your parents still support you financially, there's a sense of pride in knowing you earned your own money to spend on things you enjoy without having to ask for it. In some cases, working teens even help ease the financial burden on their families by covering personal expenses their parents would otherwise have to pay for. Besides, many young workers start setting aside money for future needs, whether for big purchases, emergency savings, or even college (Mortimer, 2009). Ultimately, consistent work experience as a teenager exposes you to real-world lessons that no amount of advice can fully prepare you for, which helps you develop a strong work ethic and smart money-management skills along the way.

Understanding Your Work Style

In addition to exploring different career paths, working as a teenager allows you to discover which work arrangements suit you best and how to protect yourself in various job settings. Whether you've worked as a cashier, restaurant server, bagger, online tutor, or pet-sitter, what conclusions have you drawn from those experiences? Do you feel more comfortable with structured eight-hour shifts, or do you prefer work that's based on specific tasks or projects? In which arrangement are you the most productive? Answering these questions can help you decide whether a traditional job or gig work is a better fit for you.

The key differences between traditional jobs and gig work come down to stability, flexibility, and the nature of the job itself. Traditional jobs typically require a long-term commitment to a single employer, with fixed hours and a set workplace. They provide a steady paycheck and often include benefits such as health insurance, retirement plans, and paid time off. On the other hand, gig work offers more control over your schedule, allowing you to choose when and where you work. These jobs often involve taking on different projects for multiple clients. However, the downside is that gig work tends to come with inconsistent income, especially when you're still building your reputation. Unlike traditional jobs, gig work doesn't offer job security, steady pay, or benefits like insurance and paid leave. So, rather than just guessing which one fits you best, consider trying both for a few months to get a clearer picture of which one is worth pursuing as your main job.

Even though both options have their pros and cons, gig work has become more accessible than ever, due to digital platforms that connect workers with on-demand jobs. Especially during economic downturns or uncertain job markets, gig work can serve as an alternative—or even a complement—to a traditional 9-to-5 job. In 2021 alone, nearly 36% of U.S. workers participated in the gig economy, and this percentage is expected to rise, potentially becoming the majority by 2027 (Ozimek, 2021). So, even if you feel that traditional employment suits you best in terms of stability and productivity, it's still worth exploring gig work as a side job. The experience can offer valuable skills, extra income, and even new career opportunities down the road.

SEARCHING FOR SIDE HUSTLE OPPORTUNITIES

While we've explored how having a job helps teens develop new skills and build confidence, many of you might still feel unsure about where to start, especially when you consider how your parents might react to the idea of you working. It's completely natural for parents to feel overwhelmed at first, whether they're concerned about your safety and academic performance or want to shield you from the pressures of adult responsibilities. So, if your parents hesitate when you ask for permission to work, remember that their concerns are likely rooted in love and a desire to protect you. The best way to ease their worries is by doing your own research beforehand. Look for age-appropriate jobs that align with your skills and interests, and be ready to explain why, where, and when you want to start working. When you present a well-thought-out plan, you'll have a better chance of gaining their support. And once they agree, show them you can handle the added responsibilities while keeping up with your schoolwork. After all, no parent wants to hold their child back from becoming independent and thriving.

Start With Local Jobs

Starting with nearby jobs for your first job is a great way to show your parents you're responsible. Neighborhood jobs can be a great starting point because they allow you to stay close to home, minimize transportation hassles, and offer more flexibility to balance work with school activities. Here are a few neighborhood jobs to consider:

- **Babysitting** is often the first job that comes to mind when teens start exploring the working world. If you enjoy caring for children or have experience looking after younger siblings, this can be a great option. It offers flexible hours while helping you develop responsibility and patience. As a babysitter, your main tasks typically include feeding, bathing, playing with the kids, and keeping an eye on them while they sleep. While it may seem simple, remember that you're responsible for another person's child, even if only for a few hours. Be sure to get clear instructions from the parents regarding routines, allergies, or special needs, and always have emergency contact information on hand. While babysitting, never leave children unattended, supervise them during play and mealtimes, and keep them away from potentially dangerous areas.

- If caring for children isn't your thing, **dog walking** could be a fun alternative. Getting outdoors while spending time with furry friends sounds like a great deal, right? Not only is it a good way to earn money, but it also helps you stay active. Besides, it requires no special equipment, and if you build up regular clients, you can make a decent hourly wage. However, you need to be sure you're confident in handling the breeds you'll be working with. The last thing you want is to be responsible for a dog that's too strong or difficult to control. Before each walk, gather all the necessary supplies, such as leashes, waste bags, water, and treats, and, of course, follow the owners' instructions.

- **Working in food service** is another solid option, especially if you want to develop customer service skills, teamwork experience, and basic food-handling knowledge. Some common entry-level food service jobs include working at fast food restaurants, serving diners, being a barista, or even scooping ice cream. While food-service jobs sometimes feel high pressure, they teach you to stay focused and multitask efficiently. However, if you don't enjoy fast-paced environments or working under pressure, this may not be the best fit for you.

- **Working in retail** can be a fast-paced experience that helps you develop problem-solving skills by handling customer complaints and resolving issues. It's a great way to improve your communication and interpersonal skills, both of which are valuable in any job. A retail job can include being a cashier, working as a sales associate, helping customers find products, or keeping shelves organized and maintaining inventory. However, keep in mind that retail jobs often require long hours of standing and occasionally lifting heavy items, so make sure you're physically comfortable with those demands. Besides that, since retail involves working closely with customers, you may encounter people who are rude or overly demanding, which can be stressful at times, so be sure you can handle that part of the job, as well.

Online Job Options

Once you've built a solid track record with in-person jobs, you can expand into online work. While traditional jobs teach you how to work with a team, handle customers face-to-face, and adapt to structured schedules, online jobs introduce an entirely different dynamic. You'll often work independently or with a virtual team, manage clients remotely, and have more flexibility in your schedule. The online job market also offers a much wider range of opportunities, many of which you might not have realized existed. Most of you already have access to digital devices like smartphones, tablets, or laptops, so why not use them to make money? Here are a few online job options to consider:

- If you enjoy writing, **freelance work for blogs and websites** could be a great fit. Platforms like Fiverr, Upwork, and Snagajob connect freelance writers with potential clients, though some sites have minimum age requirements. Whether you're writing articles, stories, or even poetry, you'll need to practice before applying for paid opportunities. Beyond making money, freelance writing helps you build a portfolio, which can be valuable for college applications and future job opportunities.

- If you have an eye for creative visuals, **graphic design** is another promising option, especially with the constant demand for social media content. Start by showcasing your designs on your social media accounts. You can create logos, illustrations, posters, flyers, and social media graphics to attract potential clients. Freelance platforms can also help you build a reputation as a designer. Most importantly, ensure that your images or fonts are properly licensed to avoid copyright issues.

- If you're strong in certain subjects, **online tutoring** is a great way to earn money while reinforcing your own knowledge. All you need is a device (phone, tablet, or laptop), an internet connection, and expertise in a subject. Tutoring not only helps your students but also strengthens your understanding of a given subject, especially if you plan to continue your education. Plus, it develops your communication and teaching skills, which can be valuable in any career.

- If you'd rather be your own boss, consider **selling your own products or services**. Whether it's baked goods, custom-made items, digital templates, or local services, you can start a business without relying on major platforms. You don't need a full website to get started—simply create a social media page dedicated to your business where customers can contact you.

No matter what online job you choose, always turn to your parents for guidance, especially when it comes to understanding job contracts, salary terms, and tax laws. Before signing up for any platform, discuss your earnings and the type of work you're doing to ensure everything meets the legal requirements and that you're not being overworked. Because online jobs come with certain risks, be cautious about sharing personal information and avoid offers that seem too good to be true.

TURNING PASSION INTO PROFIT

Once you've tried a few jobs—whether offline or online—you'll likely realize that working every day, or even occasionally, can start to feel repetitive and even boring. But does work have to be that way? What if earning money could actually be something you're excited about? That's exactly what happens when you turn your passions into profit. Monetizing your hobbies and interests isn't just a fantasy; it's entirely possible if you're willing to put in the effort. Many people have built entire careers around the things they love, and you can, too.

Think back to your favorite childhood movies—maybe *The Lion King*, *Toy Story*, or *Frozen*. These films became global sensations, generating billions of dollars. But they all trace back to one thing: Walt Disney's childhood love for drawing. No one could have predicted that his passion for art and attention to detail would lay the foundation for one of the most iconic entertainment companies in the world. And who doesn't love Ben & Jerry's ice cream? It all started with two childhood friends who shared a dislike for gym class and took a $5 ice cream–making course. What began in a renovated gas station as a simple way to make a living soon became something bigger. Etsy is another great example of how a group of people turned their passion into a thriving business. What started as a simple platform for artisans quickly grew into a $1.8 billion marketplace. Driven by their love for crafting and the DIY culture, Etsy's founders created a space that now connects 1.5 million active sellers with 22.6 million buyers (Majewski, 2015).

If they can do it, so can you. With the right approach and dedication, turning your passion into profit doesn't have to be just a dream. It starts with discovering your true passion—not what others say it should be, not what influencers claim is the most profitable, and definitely not something you're only pretending to enjoy. Passion isn't something that fades overnight; it's what consistently fuels your enthusiasm. It doesn't have to be mainstream, either—authentic passion is contagious, and simply showing how much you love what you do will naturally resonate with your audience.

Standing Out and Staying Authentic

Once you have a clear vision of your passion, the next step is finding what makes you stand out among others in the same space. Identifying your uniqueness lays the foundation for monetizing your hobby and setting yourself apart. By adding your own special twist to what you love, you can better pinpoint those who will appreciate it the most. Only then can you make a profit out of the business you're making as you fill the gap within the market. Most importantly, when starting out, authenticity is everything. Staying true to who you are—your personality, values, and voice—makes your business and brand stand out. People connect with realness, and being genuine will always be more appealing than trying to fit a mold.

Once you've established your brand values and uniqueness, the next challenge is finding your target market. Where's the best and most effective place to do that? Thanks to the digital world, it's easier than ever—people share almost everything online. However, finding customers doesn't necessarily mean chasing customers one by one, which can be exhausting. Instead, make sure your online presence speaks for itself. Your profile—where you have full control —should showcase your passion and what your brand offers.

Think about using keywords that will appeal to your target audience so that when they come across your profile, they feel compelled to follow you and explore your content. Beyond your profile, hashtags can also work in your favor—e.g., #CapturingFurAndSoul, #EcoFriendlyProduct, #LocalVeganGoodness, or #HandcraftedWithHeart. Choose some hashtags that truly reflect your brand and appeal to your ideal customers. Lastly, make it easy for people to connect with you. Provide clear contact information or ways for potential customers to reach out, ensuring they know you're available to offer products or services that appeal to them and their interests.

EXPLORING BASIC INSURANCE AND TAXES

While earning money gives you the freedom to buy what you want, it also comes with the responsibility of paying taxes. You might be thinking, *I'm just a minor—what do taxes have to do with me?* Well, just like adults, you have to pay income taxes once you reach a certain earnings threshold for the year. Below are the two main factors that determine whether you, as a minor, need to file a separate tax return from your parents.

- **Dependency status:** This refers to the rules that determine whether someone can be claimed on another person's tax return, which affects both parties' tax responsibilities. As a minor, you're generally considered a dependent on your parents' tax return, meaning you don't have to file separately unless your income exceeds certain limits. To qualify as a dependent, you must be under 19 years old (or under 24 if you're a full-time student), live with your parents for more than half the year, and not provide more than half of your own financial support (Internal Revenue Service, 2024). This rule doesn't just apply if you live with your parents. Even if you're staying with relatives or grandparents, you could still be considered a dependent if you qualify under the qualifying relative category.

- **The income threshold:** This refers to the level of earnings that determines whether you are subject to taxes. Your tax-filing requirements depend on two types of income—earned and unearned. Earned income includes wages, salaries, tips, freelance payments, and any money received in exchange for work. Meanwhile, unearned income includes money from investments, such as royalties, dividends, or capital gains. For 2024, the filing threshold for earned income is $13,850, and for unearned income is $1,250 (Internal Revenue Service, 2024). If you don't exceed both limits, you typically don't need to file a tax return.

If your earned and unearned income exceeds the thresholds, you'll need to gather the necessary documents, such as your Social Security number, W-2 forms (if you have an employer), and 1099 forms (if you're self-employed or earned investment income). From there, you can use free tax software like 1040.com, TaxSlayer, TaxAct, or ezTaxReturn.com to file your federal tax return.

Health Insurance and Other Benefits

Beyond paying taxes, having a job comes with greater responsibilities, so why not take it a step further by paying yourself back? And what better way to protect yourself and your finances than with health insurance? I know what you're probably thinking: *What could really happen to someone my age? I'm perfectly healthy, I exercise, and the worst thing I've dealt with is a cold or maybe some allergies.* But here's the thing: Illness, accidents, or sudden medical emergencies don't care about your age. You could be completely fine one moment and in need of serious medical care the next.

Health insurance doesn't just cover medical bills—it protects you from financial disaster. There are countless cases where a major illness wipes out a person's savings or, even worse, puts their family in debt. While insurance might seem expensive now, going without it could end up costing you much more in the long run. To help you get started, here are some ways you can get coverage:

- Parents' plan allows you to stay on your parents' health insurance until age 26, even if you're married, living on your own, or have a job.
- Employer plans may be offered by your employer as part of your employee benefits. This could be your most affordable option, as they typically cover part of your premium. Keep in mind that you usually need to work for a set period of time before becoming eligible.
- Medicaid provides free or low-cost health coverage for low-income individuals and families as long as they can prove their eligibility.
- Many colleges and universities offer student health plans that often cover on-campus medical services.
- Individual health insurance plans allow you to apply for your own plan, which gives you flexibility in choosing coverage. In some cases, you may be eligible for subsidies, but this type of plan is generally more expensive.

Aside from health insurance, you might need other types of coverage as well—mainly auto and renter insurance—as you plan to move out and have your own vehicle. If you drive, you're usually covered under your parents' auto insurance policy. However, once you're no longer a dependent, you'll need to get your own insurance. And when you move out and rent your own place, renter's insurance protects your belongings from theft, damage, or disasters. For most insurance, you are still currently covered under your parents' policies, but knowing your options now ensures you have adequate coverage when the time comes to transition to your own plan.

Interactive Element

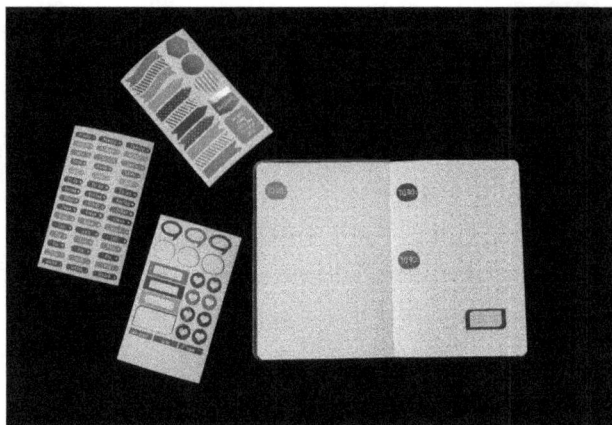

Tired of wishing you had that new phone or enough cash for pizza with your friends? It's time to level up your finances with an income goal tracker. Each month, set a target amount you want to earn and slap on a fun sticker or stamp every time you hit a mini–money milestone. Every sticker becomes a badge of progress, which is way more exciting than just staring at numbers in your bank account. Tracking your income doesn't have to be about crunching boring numbers—make it feel more like a game and less like a chore. Over time, you'll start to see patterns in how you earn, what motivates you most, and where you could maybe push a little harder. Plus, when you look back and see how far you've come— yeah, that's a confidence boost that money can buy.

WHAT'S NEXT

Earning your own money—especially from something you're passionate about—is incredibly rewarding and exciting, right? But the real challenge isn't just making money; it's keeping it and making it work for you, not against you. Many adults struggle to turn a steady income into lasting wealth, often stretching their paychecks just to get through the month, so why not start building smart money habits now? That way, when your income grows and you take on more side jobs, you won't end up losing a fortune just because you didn't have a budgeting plan. In the next chapter, we'll explore ways to design budgeting rules, track your expenses, and implement a budgeting system that you know will work best for you.

CHAPTER FOUR
R—RULE YOUR BUDGET

> *Do not save what is left after spending, but spend what is left after saving.*

— WARREN BUFFETT

Are you the type of person who saves first or saves last? Saving last means only setting aside whatever is left in your wallet—if anything—when, in reality, saving should be a priority. On the other hand, saving first ensures that you always put aside the amount you need before spending money on anything else. But how can you make sure you consistently have enough to save? That's why you need a solid budgeting plan to help you allocate your paycheck effectively, covering both your needs and wants while making sure savings remain a non-negotiable part of your financial plan. We'll discuss that in this chapter.

DESIGNING BUDGETING RULES

When you're young, earning money is mostly about covering the extras your parents won't pay for—maybe dining out with friends, grabbing those trendy shoes, going to concerts, or finally checking off items from your wish list. At this stage, managing money isn't really on your radar because you have few, if any, financial responsibilities while living under your parents' roof. It's easy to be carefree about spending, but as you transition into adulthood, that mindset needs to shift. Just like learning how to drive or cook, budgeting is just as—if not more—important because it determines how you'll afford everything else in your life. A budget doesn't just help you see where your hard-earned money is going; it ensures you have a plan for the future.

Beyond keeping your current finances stable, budgeting sets you up for long-term success and a healthier relationship with money as an adult. You've probably heard of high-earning adults still struggling with debt, running out of money before the end of the month, having little to no savings, constantly stressing about finances, or even passing down financial hardship to their kids. One study shows that more than half of those who learned about personal finance as kids have at least $5,700 in savings, compared with just 30% of those who didn't (GoHenry, 2022). Think of budgeting as a roadmap guiding you through the twists and turns of financial independence. Without it, it's easy to get lost in impulse spending and struggle to stay on track—both for now and later.

Budgeting Basics

When you first hear the word "budgeting," it might sound like a complicated concept meant only for adults. But don't worry—learning to budget as a teen can be as simple as understanding how much money comes in and how much goes out.

- **How much comes in?** If you're just starting to earn money as a teenager, figuring out how much you make is pretty straightforward, especially if you focus on one job. But as you grow older and take on more responsibilities, you may need multiple income streams to cover your needs. Or maybe you already juggle more than one job or side hustle. Knowing your total income includes your paycheck from a main job as well as earnings from side gigs, freelance work, or even a small business. Tracking all these income sources helps you see which job brings in the most money, which has the highest earning potential, and which might not be worth the effort. If you don't have a consistent monthly income yet, calculating it might feel tricky. To get a better picture, record every payment you receive—whether it's cash or a transfer—for any work you do. After tracking your earnings for three to four months, you'll have a better estimate of your average monthly income, even if it greatly varies each time.

- **How much goes out?** Have you ever found yourself wondering where all of your money went at the end of the month? You know you spent it on something, but somehow, it seems to disappear. That's a sign it's time to start tracking your expenses. Like keeping tabs on your income, monitoring your spending is just as important. It helps you see which expenses eat up most of your money and where you can cut back to reach your financial goals faster. Expense tracking also makes you more aware of impulse purchases, helping you to take control of your spending. So, start by recording every dollar you spend— what you bought and how much it cost—to help you distinguish between needs and wants.

Once you've tracked your expenses, it becomes much easier to distinguish between needs and wants. Since you're still transitioning into adulthood, you might not be paying for essentials like food, water, clothing, or housing yet—your parents likely cover those expenses. But keep in mind that in just a few years, you'll be responsible for all of it. That's why focusing your budget on saving now can give you more financial freedom when you actually need the money in the future, rather than spending it all on short-term enjoyment. That doesn't mean you can't treat yourself, but cutting back a little now can help you enjoy more later. Learning to delay gratification will help you prioritize wisely and avoid impulsive spending. Wouldn't it feel better to have money for your rent when you need it, rather than blowing it all on concert tickets today?

The 50/30/20 Rule

To make saving a habit instead of just an afterthought, you can use the 50/30/20 budgeting rule as a simple guideline:

- 50% of your income should go toward essential expenses, things you must have to function. While adults typically budget for rent, groceries, and insurance, your essential expenses might include transportation costs, school supplies, your phone bill (if you pay for it), or necessary clothing.
- 30% of your income can go toward things you enjoy— movies, gaming, dining out with friends, hobbies, or even new clothes.
- 20% of your income should be set aside for savings and an emergency fund. Since your current essential expenses likely don't take up half of your income, use the extra money to boost your savings instead of spending it all on wants.

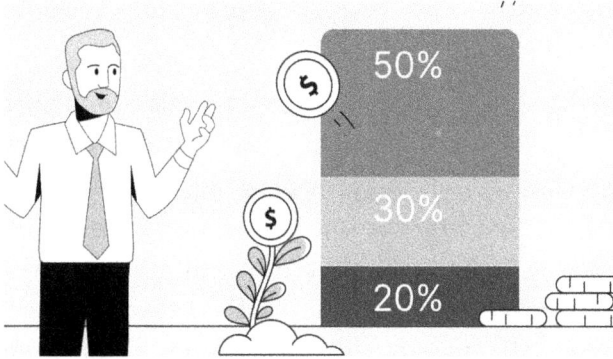

Because you're just starting with budgeting, the 50/30/20 rule makes it much easier to divide your income into categories and keep track of your spending. This simple structure also makes it easier to stick to your budget and stay on track. Even if it feels like you don't have enough money right now, just follow the rules and make it a habit. That way, when you start earning more, you won't automatically increase your lifestyle expenses. Keep in mind that earning a higher income doesn't mean that you need to spend more on luxuries. If you let your expenses grow alongside your income, you'll feel like you're getting nowhere financially, just at a different income level. Instead, keep your spending in check and focus on long-term financial security.

Take Warren Buffett, one of the richest people in the world. Despite his wealth, he doesn't splurge on expensive hobbies, luxury travel, or flashy mansions. In fact, he's lived in the same house in his hometown since 1958, even though he could afford a lavish estate anywhere (Smith, 2024). To him, wealth isn't about owning the most extravagant things—it's about buying what truly matters to you, not what society tells you to have. So, whether you're just starting out or already earning more than you need, letting your entire income slip away on unnecessary spending is the fastest way to financial downfall.

TRACKING YOUR BUDGET

Following the most effective budgeting rule won't mean much unless you take intentional and consistent steps to track what comes in and what goes out of your wallet. Guesswork and rough estimates aren't enough—you need real awareness to see whether your budget is being used efficiently and effectively. When you track your budget regularly, you gain a better understanding of your typical daily, weekly, and monthly expenses. If you notice signs of overspending, having clear records will help you pinpoint exactly when it happens, whether it's a specific day or week, or even any particular spending triggers. Especially if you ever struggle to distinguish between necessary and unnecessary expenses, consistent tracking will give you better insight into your true spending habits. It helps you to see which needs or wants are actually costing you more than you initially thought. While tracking expenses may sometimes feel like a daily battle to stay consistent, it ultimately makes financial planning much easier. Over time, it helps you identify weak spots in your spending, increasing your chances of reaching long-term financial goals with less stress. These are a few different budgeting models:

- **Simple pen and paper:** Using pen and paper is one of the simplest, yet most effective, ways to start tracking your budget, especially if you always carry them with you. If you already enjoy keeping a diary, this method can seamlessly integrate into that habit. It also offers the most flexibility—you can track your spending in whatever format works best for you, whether it's a simple list or a table with as many rows and columns as you prefer. However, if you're always online and don't usually carry a notebook, this method might make it easier to forget where your money is going and how much you've spent.

- **The envelope method:** This method takes you back to budgeting basics by dividing your cash into multiple envelopes, each assigned to a specific expense category. At the start of each month, you allocate a set amount to each category and commit to spending only what's inside the envelope. Once the cash runs out, that's it—you can't spend any more in that category until the next month. With envelope budgeting, you physically handle and monitor your budget, which makes you more aware of every expense. However, this method might tempt you to "borrow" from one category to cover another, throwing off your entire budget. Also, if you rely mostly on digital payments, the envelope method may not be the most practical option.

- **Spreadsheets:** Using a spreadsheet can help you transition from manual tracking to digital tools, as spreadsheets also use a table format. This method allows for detailed expense records, categorization, and, most importantly, financial analysis. Unlike manual tracking, spreadsheets offer more flexibility and make it easier to generate weekly, monthly, or even annual reports to better understand your spending patterns. If you've already used spreadsheet apps for school or work, this method will be a breeze, especially since most spreadsheet tools are free and accessible (e.g., Google Sheets) as long as you have an internet connection. Besides, working with spreadsheets helps you build useful skills in data entry, organization, and basic financial calculations— valuable for both personal and professional use. However, learning to use formulas and advanced settings can be a bit challenging if you've never used them before.

- **Budgeting apps:** Mobile apps are likely the most convenient tracking method—after all, you're already glued to your phone every day. With just a few taps, you can record expenses as they happen, reducing the chances of forgetting purchases and giving you a more accurate view of your spending habits. Most apps also let you set limits and goals for specific categories, helping you to stay on track. Some apps even integrate with your bank account, syncing transactions automatically so you don't have to check both the app and your account manually. To get started, you can try You Need a Budget (YNAB) or EveryDollar, both of which focus on zero-based budgeting, ensuring that every dollar is assigned to a specific expense or savings goal. Or, you can try PocketGuard, which offers an "In My Pocket" tab, showing exactly how much money you have left to spend after bills and other necessary expenses. Give yourself time to experiment with different apps to see which features work best for you and your financial goals.

IMPLEMENTING A BUDGETING SYSTEM

In addition to choosing a budgeting method and tracking your expenses, consistency is the key to long-term financial success. Staying consistent might sound like the simplest advice you've heard, but trust me—it's not always easy to stick with it. It takes strong motivation and a series of intentional steps to build a system that eventually becomes second nature, something you do without even thinking about it, like any ingrained habit. Think about the habits you've formed over the years, whether it's reading, exercising, or socializing. These patterns were likely influenced by your family and inner circle, and over time, they became automatic. The same can happen with your financial habits.

Now is the perfect time to build a similar system for your personal finances, especially if you're just starting to earn and manage your own money. In fact, without even realizing it, you may already have financial habits that were influenced by your parents—whether it was running a small lemonade stand, selling items at a garage sale, or saving part of your allowance in a piggy bank. And who knows? Maybe that childhood experience of selling lemonade sparked a passion for entrepreneurship, leading you to sell baked goods, crafts, or even start a business as an adult. Likewise, the habit of saving coins in a piggy bank can develop into a habit where you are saving hundreds of dollars a month, or even automating your savings as an adult. The key is to choose the financial systems and habits that will serve you best in the future. Even if you're starting small now, you're planting the seeds of consistency. As your income grows and your financial knowledge expands, those positive habits will naturally evolve to work in your favor. So, don't think of budgeting as a quick sprint—it's a marathon. Focus on developing consistent habits, gradually advancing your strategies, and building long-term financial stability.

Interactive Element

Category	Projected Budget	Actual Spending	Difference -/+
Income			
Allowance			
Birthday Gifts			
Part-Time Job			
Gig Work			
Total Income			
Expenses			
Cell Phone			
Clothes / Shoes			
Donations			
Eating Out			
Entertainment			
Hobbies			
Personal Care			
Presents			
Savings			
Snacks / Coffee			
Total Expenses			
Net Balance			

Creating a budgeting worksheet is your secret weapon for finally ditching the constant question of "Where did all of my money go?" at the end of the month. When setting up your budget table, include three sections—one for your projected budget, one for your actual spending and one for the difference. Label the categories based on your typical expenses—"Needs," "Wants," "Hobbies," or anything else that fits your lifestyle. In the "Projected" section, estimate how much you think you'll spend in each category. Then, throughout the month, track your actual spending and write down exactly what you spent in the "Actual" table. At the end of the month, compare your projected budget with your actual spending, and the difference between them to see where you may need to make adjustments.

WHAT'S NEXT

Once you've organized your money and become more serious about your work performance, know that working hard isn't the end goal. What if I told you that your smart spending habits could help your money to increase, even when you're not actively working? Well get ready, because in the next chapter, you'll discover how your savings have the potential to multiply into serious cash for the future while still allowing you to enjoy the present.

CHAPTER FIVE

T—TREASURE GROWTH

> *The habit of saving is itself an education; it fosters every virtue, teaches self-denial, cultivates the sense of order, and trains to forethought.*

> — T.T. MUNGER

Saving little by little for your needs and wants can feel like a tiny, never-ending task, especially when you're just starting out. But remind yourself how incredibly fulfilling and rewarding it is when you finally get to buy something you've wanted for a long time. Remember that feeling of pride? And how your parents probably felt just as happy and proud of you, too? Now imagine feeling that kind of confidence all the time, knowing you're in control of your money and ready for whatever life throws your way. In this chapter, we'll explore simple ways to start building your treasury today.

MULTIPLYING YOUR SAVINGS

Before you even think about growing your savings, let's start with the basics—having some savings in the first place. Growing up, you probably heard your parents talk about saving all the time, whether it was for movie tickets, concert tickets, trendy clothes, or other purchases your parents wouldn't pay for. As you transition into adulthood, the need to have savings is much more important as you'll be facing bigger expenses like rent, a car, computer, or even buying your own home. And, trust me—your future self will thank you for easing the stress of wondering whether you'll have enough money for those major purchases.

While saving money doesn't really have a downside, many people —especially teens and young adults—hesitate to save money because it can feel unrewarding in the short term. That mindset might make you avoid it altogether. But what if I told you that saving doesn't have to start with hundreds of dollars? In fact, you can start with less than a dollar, a process known as *micro-saving*. Rather than requiring you to save hundreds of dollars at once, micro-saving relies on your consistency, even with the smallest amounts.

Micro-Saving

Micro-saving is a simple financial strategy that lets you set aside small amounts of money, no matter how tiny, into your savings. Think of saving like a marathon—micro-saving is your daily training run. Just like any habit, it starts small before it becomes second nature. Micro-saving also helps you become more aware of seemingly minor spending habits that add up over time. Maybe you grab a cup of coffee twice a day—what if you cut back to just one? Or maybe you're paying for subscriptions you barely use. By recognizing these small, unnecessary expenses, you can redirect that money toward savings instead. You'd be surprised how much those little adjustments add up, without really having to change your lifestyle very much.

Once you've identified these spending patterns, you can start collecting spare change or small bills in a jar somewhere visible. Watching it grow over time can be surprisingly motivating. But if you prefer digital transactions, no worries—some apps, like Acorns, Chime, and Qapital, automate the process for you, which helps round up your transactions to the nearest dollar and deposit the difference into your savings. If you buy a slice of cake for $4.75, the app rounds it up to $5.00 and saves the extra $0.25. No manual calculations are needed, and there's little to no fees.

Pay Yourself First

To make sure you never forget to set money aside for savings, consider trying the pay-yourself-first method. As the name suggests, this approach means setting aside a portion of your income for savings immediately after getting paid—before spending a dime on bills or discretionary purchases. This strategy is particularly beneficial at this stage in your life, before other financial commitments arise. It helps you save more by prioritizing savings over spending, rather than just putting aside whatever is left after covering your needs and wants. To get started, decide how much of your income you want to save and what you're saving for. Then, set up an automatic transfer to your savings account each time you get paid. Once your savings are secured, you can manage what's left for your essential and non-essential expenses.

Compound Interest

Still hesitant to start saving? Maybe you think you're still young and don't have much to worry about. But that's exactly why you should start now! With fewer financial commitments, this is the perfect time to take advantage of compound interest, one of the most powerful tools for growing your money. Compound interest lets your money work for you instead of the other way around. It allows your savings to earn interest, which then gets added back to your principal. Over time, each new interest calculation is based on an increasingly larger balance, making your money grow faster and faster. Think of compound interest like a snowball rolling downhill —starting small but growing bigger and bigger as it picks up speed. The sooner you start, the longer your money has to grow. By putting your savings into a high-yield savings account, you maximize this effect, letting your money multiply exponentially over the years.

ACHIEVING FINANCIAL GOALS

After a while, consistently contributing to your savings may feel repetitive and boring, making it tempting to stop altogether. That's why you need to have a strong motive and clear goals behind your saving habits. Without goals, you're essentially walking without direction, and before long, you might give up, leaving your financial future unclear. Setting goals makes saving more exciting and turns your financial journey into a purposeful mission. When you have a clear destination, you become more mindful and intentional with your money, ensuring every dollar is used wisely. Think of it like having a GPS for your financial success—it gives you direction, purpose, and, most importantly, a solid reason to push through, even when you don't feel like saving. So, stop merely surviving financially—it's time to thrive.

SMART Goals

Take a moment to anchor yourself by implementing the SMART goals described below. SMART is an acronym for Specific, Measurable, Achievable, Relevant, and Time-Bound goals.

- **Specific:** Vague goals won't get you anywhere—they'll only leave you confused about your financial decisions. Instead of saying, "I want to be rich" or "I want financial freedom," define what that actually means for you. Ask yourself: *What does being financially secure look like for me? What specific items or experiences do I want to save for?* or *What does being good with money mean to me?* Answering these questions helps you gain a better understanding of your financial goals. It guides you toward more specific ones, such as *I want to save $1,000 for a new phone or laptop," I want to save $200 for a concert,* and *I want to build a three-month emergency fund.* Don't these clear, specific goals feel more motivating and exciting? Just like you spend time carefully planning a vacation, your finances deserve the same level of clarity.
- **Measurable:** Along with being specific, your goals should be measurable so you can track your progress. Measurable goals keep you accountable and motivated, ensuring you stay on track. They also help you assess whether you've truly reached your target. Instead of saying, "I want to be rich" or "I want to save more," set a concrete goal like *I will*

save $100 each month or *I will build an emergency fund covering three months of expenses.*

- **Achievable:** Dreaming big is great, but your goals should also be realistic based on your current financial situation. Ask yourself, *Is this goal attainable with my current income and expenses?* and *Have I already made progress toward this goal, or am I starting from scratch?* If you've been saving for a few months already, you can start setting more challenging goals. But if you're starting from zero, it's better to begin with smaller, achievable milestones to build momentum.

- **Relevant:** When it comes to financial goals, most adults seem to chase the same dreams—getting a car, buying their first house, or planning a big wedding. But it doesn't have to look like that for you. Your financial goals should reflect where you are in life right now, what you might need in the future, and what's actually within your reach. Instead of going along with what everyone else thinks you should want, focus on what you value. If you're not into trendy, fast fashion, then there's no reason to make it one of your savings goals. Maybe you'd rather spend more on your hobbies, and that's perfectly fine. Or if the idea of paying for gas and car maintenance doesn't fit your budget (especially if you live somewhere with decent public transportation), then it's totally okay to hold off on saving for a car. Whatever goals you're setting, they're valid. You don't need anyone's approval for what you choose to save for.

- **Time-Bound:** Most importantly, you need to give each goal a clear timeline—short-term, medium-term, or long-term. Adding a time frame makes your goals feel more real and keeps you accountable. Without a time frame, goals can become vague wishes that get pushed aside for more urgent stuff. When setting those timelines, aim for a balance between realistic and challenging. You don't want

your goals to feel so difficult that they discourage you—or
so easy that they don't push you to grow.

When it comes to setting goals and making sure they stick, it all
boils down to your everyday spending habits. You always need to
be aware of where your money is going. Take a moment to review
your current spending patterns and the stuff you already own.
Look around your room or your home. Do you feel like your closet
is overflowing with clothes you rarely wear? Have you been going
to concerts, not even knowing who's performing? Or are you
stocking up your pantry only to find things expired at the end of
the month? If any of that sounds familiar, there's a good chance
you're spending way beyond what you actually need—or even
truly want—in a lot of areas. So, from this point on, be intentional
with every purchase. Having a mindset of gratitude for what you
already have is key to reaching your goals while finding a balance
between needs and wants.

Keeping Spending Triggers at Bay

Even when you're trying to be mindful of your spending, it's easy
to slip up. Ads from online stores pop up, or a quick trip to the
mall turns into a shopping spree. Whatever your triggers are,
know that these things are designed to make you want more, not
to help you figure out what you really need. So, do yourself a
favor: Unsubscribe from retail promo emails, skip the commer-
cials, and hang out somewhere other than the mall. Breaking these
habits can go a long way in helping you avoid impulse buys that
only give you temporary satisfaction. If you're ready to take it a
step further and embrace a more minimalist lifestyle, try this trick:
Every time you bring in something new, get rid of a similar item.
Bought new shoes? Donate an old pair. In the beginning, it might
be easy if you've got a lot to work with. But as your wardrobe gets
smaller, you'll start being more thoughtful about what's truly

worth keeping—what still fits your style, your life, and your goals.

BUILDING AN EMERGENCY FUND

Having an emergency fund is just as important as saving for your needs and wants. Just like adults, you're not immune to unexpected expenses that can wipe out your savings in an instant. Think about it: Needing to fix your car, losing your phone, getting your bike stolen, or breaking your glasses are things that might seem small, but they can be surprisingly expensive to deal with. Especially if you're still relying on part-time or seasonal jobs, where paychecks come and go, this is the perfect time to start building an emergency fund. Without one, you might feel tempted to reach for a credit card —which, trust me, isn't a great habit to start at a young age—or borrow money from friends. Sure, it might feel like a quick fix during a crisis, but it can quickly turn into a pile of debt that's hard to pay back.

Learning to set aside money for emergencies helps prepare you financially, giving you a safety net for the unexpected. It also builds good habits that will serve you well later in life, when your responsibilities and expenses increase. So, how much should you aim for in your emergency fund? The more, the better—especially for bigger or longer-term financial emergencies. But as a starting point, since most of your living expenses are probably still covered by your parents, aim for something between $500 to $1,000. As your income becomes more stable and you take on more financial responsibility, work toward building up 3–6 months' worth of living expenses. Setting a specific goal for your emergency fund isn't meant to overwhelm you or make you feel discouraged about how far you still have to go—it's about having a clear target. Even saving a dollar or two at a time counts and adds up.

Opening a Separate Savings Account

Once you get in the habit of saving and want to stay on track, setting up a separate savings account for your emergency fund can be a smart way to avoid the temptation to dip into it for non-emergencies. A dedicated account also lets you watch your savings grow over time without mixing it in with your everyday spending. However, having a separate account might mean juggling multiple minimum balance requirements, depending on the bank, so if your income is still limited or inconsistent, managing a dedicated emergency fund account might be tough. But if maintaining the minimum balance isn't a problem, start exploring the right type of account for your emergency savings. Would a basic savings account be a good fit—something with little to no fees and low minimum balance requirements, specially designed for teens? Or would an online savings account work better—one that often offers slightly higher interest rates to help your money grow? Whatever you choose, make sure to compare different banks and account options. Look closely at fees, interest rates, ease of access, and—if needed—parental controls.

Interactive Element

To budget your money more effectively, consider your spending in terms of three simple buckets: Everyday Spending, Emergency Savings, and Goal Savings. Under each bucket, write down how much money you plan to set aside. Then, below that, list out the kinds of things you might use that money for. For Everyday Spending, you might include stuff like snacks at school, bus fare, or school supplies. Under Emergency Savings, write down things like fixing a flat tire on your bike, replacing the cracked screen on your phone, or replacing something you lost. For Goal Savings, you could include saving up for a concert ticket, a new pair of shoes, or upgrading your phone or other gadgets.

Everyday Spending	Emergency Savings	Goal Savings
$40 / Month	$20 / Month	$30 / Month
Snacks at School	Bike Repair	Concert Ticket(s)
Bus Fare	Replacing a Backpack	New Pair of Shoes
School Supplies	Fixing a Cracked Screen	Upgrading Tech Gadgets

WHAT'S NEXT

You've learned how to help your money to grow even when you're relaxing, so now, let's make sure you're spending that gold wisely. Being smart with your money isn't just about pinching every penny; it's about knowing when and how to spend it right. If you've started earning a good amount—whether from your job, side hustle, or both—it's easy to get tempted to splurge on things that feel like needs but might not actually be. In the next chapter, we'll talk about how to tell the difference between needs and wants and how to make those decisions for yourself without feeling pressured or caught up in the wave of FOMO.

CHAPTER SIX

M—MAKE SMART SPENDING CHOICES

> *Most people spend money they haven't earned to buy things they don't want to impress people they don't like.*

— WILL ROGERS

Have you ever wanted to buy something so badly that you didn't even think twice, only to regret it later? Maybe it was a pair of trendy shoes or a game you didn't even play much after buying it. Spending can feel great at the moment, but smart spending makes that feeling last. In this chapter, we will explore how you can make more mindful purchases and not fall into the pitfalls of impulse shopping. We'll look at ways to balance your wants and needs, get the most out of your purchases, and stay safe from marketing stunts and peer pressure. When it's all said and done, you can enjoy your money and make choices that serve your personal goals.

NEEDS VS. WANTS

One of the most valuable skills in personal finance is learning to distinguish between necessities (needs) and desires (wants). Indeed, spending is a fact of life, but the things we spend money on aren't all alike. Needs are the basics of life, including a safe place to sleep, food to consume, and shoes to keep your feet from becoming dirty. Meanwhile, wants are the luxuries that might spice up life a little but aren't strictly essential. Think about your go-to pair of shoes. You might be eyeballing a new pair of shoes—usually a want —but your trusty old kicks still get the job done, proving they meet your real needs. The ability to distinguish between needs and wants is a key component of financial literacy, allowing you to make informed decisions.

A great way to start making better decisions between needs and wants is to ask yourself a few simple questions before making a purchase:

- *Do I really need this?*
- *Will I actually use it, and will it make a real difference to my life?*
- *Can I pay for it without compromising my other priorities?*

Aside from answering those questions, another helpful way to decide whether or not to make a purchase is by calculating the cost per use. To what extent will you really put this item to use? If you buy a pair of shoes for $50 and wear them 50 times, each wear costs $1; if you buy a $20 blouse and wear it once, it costs $20 per wear. Sometimes, a purchase might feel expensive at first, but figuring out the cost per use can help you see the real value. It gives you a better sense of how long an item will last compared to similar items at different price points. It also helps you avoid getting swept up in trends or impulse buys by shifting your focus to long-term use and actual wear. No more letting things sit around and collect dust.

Calculating the cost per use helps you to see the real value of your purchases and to make wiser long-lasting, cost-effective choices.

To start calculating the cost per use, take the price of an item and divide it by how often you expect to use it. A $100 jacket that will last for years and be worn often is much more cost-effective than a $30 jacket that will last only a season. The beauty of these formulas is that by computing cost per use, you can make better decisions and avoid wasting money on things that won't do you well in the long run.

SETTING SPENDING BOUNDARIES

Taking charge of your spending choices means being forthright about how much money you spend and being honest about it. Knowing how to manage funds properly is essential, regardless of whether you receive an allowance or work a part-time job. Making better decisions is as simple as keeping track of your spending and understanding where your money goes. For instance, you might modify your spending habits and set limits on fast food and online shopping if you are aware that you tend to go overboard on such expenses. Being honest with yourself and making choices that get you closer to your financial goals are the most important things.

Taking control of your spending starts with establishing spending limits and being responsible for them. Being frugal today isn't the only thing that matters; setting spending limits will set you up for future success. Your future spending patterns will be shaped by the decisions you make now. So, get going right away, keep tabs on your expenses, and figure out how to put every dollar to good use. By doing so, you actively lay the groundwork for financial freedom while also avoiding financial hardship.

Mindful Spending

Without limits in place, it's easy to overspend, especially with online shopping as the magnet of impulse spending in today's world. With ads blasting your feed, supposedly time-sensitive deals, and promos by influencers, it can be easy to get caught up in the frenzy of buying something. But there are things you can do to protect yourself from these temptations. A great way to do this is to practice mindfulness in your spending. Now, before you click "Buy," consider whether you're actually in control of the decision or whether the advertisement is tugging you along. You could also experiment with the 24-hour rule: Hold off for a day before buying something you don't need. That gives you time to decide whether the purchase is worth it. Moreover, reducing your marketing exposure from phone, email, or social media can help you resist temptation. So, unsubscribe from email lists, unfollow people on social media who encourage mindless spending, and avoid going on sites that make you buy things on impulse. Or, if credit cards make you lose track of your spending limits, you can switch to cash instead.

DEALING WITH PEER PRESSURE AND FOMO

The term *FOMO* has become more common as social media has become more prevalent. As people scroll through posts showing off luxuries, new trends, and the latest must-haves, it's easy to feel like you're missing out. FOMO—short for *fear of missing out*—creates this pressure to have or experience what others already seem to enjoy. At first, it might seem harmless. But for many—especially teens who are still figuring out who they are—it can trigger anxiety and make them focus too much on material things. Instead of discovering their own values or interests, they start chasing what influencers or peers show online. Sadly, this often leads young people to fall into debt—sometimes even more than adults—because they haven't fully developed self-control. A recent study also found that low self-regulation in digital environments can fuel online shopping addictions, leading to poor money habits (Nyrhinen et al., 2023). As it turns out, FOMO isn't just a feeling; it can be a serious trigger for financial problems.

So, how do we keep FOMO from taking over mentally and financially? It starts with realizing a simple truth: There will *always* be things you can't have or experiences you can't be part of. Whether it's because of financial limitations, family values, or personal choices, you won't always get what others have—and that's okay. Maybe you've always wanted blonde hair but were born a brunette. You might not be allowed to dye your hair yet, or you simply can't afford the salon visit. Either way, some things can wait. And sometimes, with a little time to think about it, you realize that what you thought you wanted wasn't really your thing after all.

Step Away and Clear Your Head

If you catch yourself getting anxious or making impulsive decisions after scrolling through social media, it might be time to unplug for a while. If your phone is the first thing you check when you wake up, when you're in the bathroom, or during every spare moment—you may be too wrapped up in the digital world. Stepping away gives you space to breathe and reconnect with yourself without the constant comparisons. It clears your head from the negativity we don't even realize we're soaking in. We get so caught up in the next big thing—the trending clothes, the latest creator contest, the newest phone or gadget—that we stop enjoying the simple, everyday joys that don't cost anything.

So, next time you're not on your phone, take it slow. Savor the small stuff. That cup of coffee, the walk home from school, the time spent playing with your siblings—these are the moments that truly matter.

Interactive Element

Give yourself a week to log your spending. Keep track of *everything* you buy, and write your spend as a "want" or "need." At the end of the week, review what you bought and look for patterns. Were there areas you overspent on wants? Did you make any impulse purchases? Doing this exercise will help you reflect on where you can improve and see how you can be a wiser spender going forward.

WHAT'S NEXT

Now that you've got the knowledge and tools, you're ready to make smarter choices with your money. Again, your goal is not to cut out all spending or never let yourself enjoy the things you want. The goal is to spend wisely based on your values and long-term goals. With some planning, you can treat yourself *and* save for the future. Smart spending sets the foundation, but now it's time to level up by learning about credit and debt—not as something to fear, but as tools that can work in your favor when used responsibly.

CHAPTER SEVEN

O—OWN YOUR CREDIT AND DEBT

> *Before borrowing money from a friend, decide which you need more.*
>
> — ADDISON H. HALLOCK

How can credit be either your worst nightmare or your greatest superpower? Honestly, it all depends on how you use it. The fear of ruining your financial future with debt isn't solved by simply avoiding credit altogether. Trust me—hiding from credit cards and loans won't prepare you for adult life. What truly makes the biggest impact is understanding how credit works and using it to your advantage. No matter where you're starting from—whether you've never had a credit card or you're already managing student loans—you already have what it takes to build a strong credit history. So, let's dive into practical ways to make credit work for you starting today.

MASTERING THE BASICS OF CREDIT CARDS

Credit and debt are two sides of the same coin. Credit is the ability to borrow money with a promise to repay it later, while debt is what you owe after you've borrowed. Understanding this relationship is the first step toward financial freedom. Credit cards might seem magical—swipe now, pay later—but there's nothing magical about the interest rates if you don't pay your balance in full. When you use a credit card, you're borrowing money from the card issuer with an agreement to repay it, usually within a month. If you don't repay the full amount, you'll start accruing interest, often 15–25%. That $50 dinner could end up costing you $70 if you let it sit on your credit card for months.

You need to understand this concept of borrowing and repaying in order to understand credit. Every time you use your credit card, you're essentially taking out a mini-loan. The credit card company pays the merchant immediately, and you promise to repay the credit card company. This convenience comes with responsibility—tracking your spending and ensuring you can repay what you've borrowed.

Most credit cards provide a grace period—typically 21–25 days after your billing cycle ends—allowing you to pay your balance without accruing interest. This is where the real power of credit cards emerge. If you regularly pay your balance in full within the grace period, you're essentially getting an interest-free short-term loan every month, plus the benefits of building a credit history and potentially earning rewards.

Speaking of rewards, they represent another dimension of credit card usage that can work to your advantage when managed properly. Many cards offer cashback, points, or miles for your purchases. These rewards can translate to savings on travel, statement credits, or gift cards. However, the value of these rewards quickly diminishes if you're paying interest on carried balances. For instance, a credit card offering 2% cash back but charging 18% interest will put you at a significant net loss if you don't pay in full.

The Best Credit Card for You

These are credit cards that can help you decide the right tool for your financial situation:

- **Secured credit cards:** These require a security deposit that typically becomes your credit limit; they're excellent for building credit from scratch.
- **Student credit cards:** Student credit cards are designed with younger users in mind, often featuring lower credit limits and educational resources.
- **Rewards cards:** These popular credit cards offer points, miles, or cash back on purchases, but can carry higher interest rates or annual fees.
- **Balance-transfer cards:** These cards allow you to move debt from one card to another, often with a promotional 0%

interest period (for a limited amount of time), which can be useful for debt consolidation.

Calculating Your Credit Score

No matter what type of credit card you choose, how you use it matters most, as it can either boost or harm your credit score. Think of your credit score as a report card for your finances. It ranges from 300 to 850, and higher scores indicate greater trustworthiness. Since payment history accounts for 35% of your score, which is the most significant factor, make sure to pay bills on time consistently. Even when you cannot pay in full, aim for at least a minimum amount. Credit utilization—the calculation of how much of your available credit you're using—accounts for another 30%. In other words, keeping your balances low relative to your credit limits can significantly boost your score.

The remaining factors in your credit score calculation include the length of your credit history (15%), new credit applications (10%), and the credit mix (10%). The length of your credit history takes into account both the age of your oldest account and the average age of all your accounts, which is why closing old accounts may sometimes hurt your score. Although, a new credit application might help boost your score by improving your credit mix or lowering your utilization, they can also temporarily lower your score due to "hard inquiries" as well. So, it's wise to space out credit applications. Lastly, *credit mix* refers to the variety of credit accounts you manage—credit cards, retail accounts, installment loans, mortgage loans, and so on. A diverse mix demonstrates to lenders that you can manage different types of credit responsibly.

Because your credit report contains the information used to calculate your score, you're entitled to one free credit report annually from each of the three major credit bureaus—Equifax, Experian, and TransUnion—through AnnualCreditReport.com. Reviewing these reports helps you catch errors or fraudulent activity that could damage your score. If you find inaccuracies, you can dispute them directly with the credit bureau, which is legally obligated to investigate and correct verified errors.

When evaluating creditworthiness, lenders consider factors other than the credit score, such as income, employment history, and debt-to-income ratio. If you haven't heard about the debt-to-income ratio, it is essentially the total of the monthly debt payments divided by the gross monthly income. The ratio ideally should remain below 36%, with no more than 28% going toward housing costs. A lower ratio indicates sufficient income to take on additional debt responsibly.

Why Good Credit Matters

Most people might say good credit opens doors. The impact of good credit extends far beyond financial transactions. For example, landlords often check credit reports when evaluating rental applications, so a strong credit history can be the difference between securing your dream apartment and settling for less desirable housing. Also, insurance companies in many states use credit-based insurance scores to set premiums, with lower scores resulting in higher rates. Within your job scope, some employers might review credit reports (with permission) for positions involving financial responsibility, viewing poor credit as a potential risk factor. Even utility companies sometimes require security deposits from customers with limited or poor credit history.

To illustrate the long-term financial impact of good credit, consider the example of two friends purchasing identical $250,000 homes, each with a 30-year mortgage. The friend with excellent credit (750+) might secure a 4.5% interest rate, resulting in a monthly payment of around $1,267 and a total interest of $206,132 over the loan's life. The friend with fair credit (620–639) might likely receive a 6% rate, creating a monthly payment of $1,499 and a total interest of $289,595. That credit score difference costs the second friend an extra $83,463 over 30 years—money that could have funded retirement, education, or other financial goals.

MANAGING DEBT LIKE A PRO

People fall into debt for various reasons—unexpected medical bills, losing a job, or sudden accidents. Sometimes, it's simply lifestyle creep—spending more as you earn more without increasing your savings. Whatever the cause, getting out of debt starts with one thing: awareness and a plan.

Paying off debt is about both the numbers and the mindset. Debt doesn't just affect your bank account. It can take a toll on your mental health, increasing anxiety, depression, and even causing physical issues (American Public Health Association, 2021). That's why ignoring your debt or falling behind on payments is the last thing you want to do—it only makes things worse. Start by writing down everything you owe—how much, to whom, the interest rates, and the minimum payments. This gives you a clear picture and sets the stage for a solid debt repayment plan.

Tackling Debt

When tackling debt, consider two popular strategies: the avalanche method and the snowball method. The avalanche method focuses on paying off high-interest debts first, saving you money in the long run, while the snowball method targets your smallest debts first, giving you quick wins to build momentum. Both can be effective—choose the one that better matches your personality and motivation style.

The Avalanche Method

With the avalanche method, you make minimum payments on all your debts and put any extra money toward the one with the highest interest rate. This approach is the most cost-effective because it reduces the amount of interest paid over time. Once you eliminate the highest-interest debt, you move on to the next one, creating an "avalanche" of payments as you progress. If you have a $3,000 credit card at 22% interest, a $5,000 car loan at 7%, and $20,000 in student loans at 5%, you would focus on the credit card first—even though it has the smallest balance—due to its high interest rate.

The Snowball Method

With the snowball method, you still make minimum payments on everything, but put any extra money toward your smallest balance, no matter the interest rate. As you wipe out each small debt, the wins give you a psychological boost that keeps you motivated. Using the same case above, you'd still start with the $3,000 credit card, but you'd go after the $5,000 car loan next and finish with the student loans.

Debt Consolidation

Sometimes, though, you might need more than just a snowball or an avalanche. If your debt is spread across too many accounts or your interest rates are sky-high, consider debt consolidation. This means combining your debts into one loan—ideally with a lower interest rate—to simplify payments and possibly save money. Options include balance-transfer credit cards, personal loans, home equity loans, or student loan consolidation. You could also look into debt-management plans, which are typically set up through nonprofit credit counselors. They work with your creditors to lower interest rates or waive fees, but they often require you to close your credit accounts. In more extreme cases, debt settlement (negotiating to pay less than you owe) or bankruptcy may be necessary, though both can seriously hurt your credit and should be a last resort.

Debt Repayment Plans

Once you've listed your debts and picked your strategy, it's time to create your personal repayment plan. Figure out how much extra money you can put toward your debt each month—even just $50 can make a big difference over time. Start by calculating your debt-to-income ratio—how much of your income goes toward debt payments. Then, review your spending to find areas where you can cut back: unused subscriptions, too much takeout, or impulse buys you could skip. Many people also use "found money" to pay off debt faster—things like tax refunds, bonuses, gifts, or extra income from a side hustle. These windfalls can help without changing your regular budget.

Another key part of managing debt is talking to your creditors. If you're falling behind, reach out before things get worse. You might qualify for hardship programs, interest reductions, or flexible payment plans. Most creditors would rather work with you than send your account to collections. If you have student loans, look into income-driven repayment plans, deferment, or forbearance during tough times. These can give you breathing room without defaulting.

Your debt repayment strategy should also include both immediate actions and long-term habits. Consider setting up automatic minimum payments on all debts to prevent late fees and credit damage, then manually making additional payments toward your target debt. Create visual representations of your progress—debt thermometers, spreadsheet charts, or mobile app trackers—to maintain motivation during the months or years of repayment. Celebrate meaningful milestones like paying off individual debts or reaching specific reduction percentages, but choose rewards that don't involve taking on new debt.

While you're focused on paying off debt, don't forget to build an emergency fund. Without one, any surprise expense—car trouble, medical bill, or lost income—can send you right back into debt. Even a small cushion of $500–$1,000 can make a huge difference and keep you on track.

Steering Clear of Bad Offers

Finally, staying out of debt is just as important as paying it off. That means being aware of offers that seem helpful but can quickly trap you in deeper financial trouble:

- Payday loans often come with insane fees, equivalent to annual interest rates of 300–700%. The Consumer Financial Protection Bureau reports that more than 80% of these loans get rolled over or renewed, creating a cycle that's hard to escape (CFPB, 2014).
- Auto title loans use your car as collateral. If you miss a payment, you risk losing your vehicle—plus, the interest rates are typically around 300% annually.
- Rent-to-own deals for furniture or electronics might seem manageable, but you'll usually pay 2–3 times more than the item's retail price.

- Store credit cards, while not as risky, often come with interest rates of 25–30%. They also offer perks that encourage you to keep shopping at that store, even when it's not the best deal.

IMPLEMENTING SMART BORROWING

Not all debt is created equal. Borrowing money to invest in things that grow in value or increase your earning potential—like education, a home, or starting a business—is often considered "good debt." On the other hand, taking out loans for things that lose value quickly—like a vacation or the latest smartphone—typically falls into the "bad debt" category.

You might have heard that having debt helps build credit, and while that's partly true, it doesn't mean you need to carry a mountain of debt. You can start building a solid credit history with just a small secured credit card. The key is to make small purchases and pay them off right away to show you're a responsible borrower.

Good Debt vs. Bad Debt

The difference between good and bad debt isn't just about what you're buying. Good debt usually comes with lower interest rates, potential tax benefits, and the chance to grow in value or boost your income. For example, mortgage interest is often tax-deductible, and homeownership can build equity over time. Student loans—while overwhelming for many—can increase your earning power if used to gain in-demand skills or degrees. Business loans can generate income far beyond their cost.

Bad debt, in contrast, tends to fund things that lose value and don't pay you back. You might need a car loan for commuting, but a vehicle loses 15–25% of its value in the first year and keeps depreciating. Using credit cards for things like takeout, entertainment, or clothes means you're often paying extra in interest for something that's already been used. And payday loans or high-interest personal loans for non-essential spending are among the worst types of debt—high cost, no return.

So, being a smart borrower doesn't mean avoiding all debt. It means being intentional. Ask yourself:

- *Will this purchase hold its value or increase over time?*
- *Will it generate income or help me earn more in the future?*
- *Is the interest rate fair, considering what I'm getting out of it?*
- *Could I just save up instead of financing it?*
- *Will this debt hold me back from reaching other financial goals?*

Along with these questions, it's also important to know how much debt you can handle based on your income—consider more than just your paycheck but any passive income that could help with debt payments. A good rule of thumb is to keep your total monthly debt payments under 20% of your take-home pay. That way, you still have room to save, invest, and enjoy life without feeling buried.

Beyond considering your current income, evaluating your income stability and growth potential provides an important context for borrowing decisions. How stable is your job? Are there opportunities for raises or promotions? How in-demand are your skills in today's job market? Someone in a more unpredictable industry might want to keep a lower level of debt, while someone on a clear upward career path may have more flexibility. If you're early in your career and expect your income to rise, you might take on more debt temporarily. But if you're already at your peak earning years, it's usually best to stay conservative.

Smart borrowers also consider interest rate environments and economic conditions. When interest rates are low, locking in a fixed-rate mortgage or student loan can be a smart long-term move. But taking on variable-rate debt (debt with rates that can change with the economy) during low-rate periods is risky—those payments can spike when rates rise. Paying attention to the bigger financial picture can help you time major borrowing decisions more wisely, like buying a home during a market dip or going back to school during a recession, when job opportunities are limited anyway.

At the end of the day, credit is a tool, not a trap. Used wisely, it can accelerate your path to financial success. Build credit by paying bills on time and borrowing only for things that add value to your life, things that grow your wealth or give you real benefits. Make sure you know where your income is coming from and stick to a budget that includes debt payments. Don't borrow just to keep up appearances. Borrow to invest in yourself and your future.

Interactive Element

Having credit and managing debt are major signs that you're stepping into adulthood—and they come with real responsibilities. Why? Because how you handle credit affects your entire financial future. Before you even think about getting a credit card or taking out a loan, go through this Credit Health Checklist. It's designed to help you identify what you already know and what you might still need to learn about using credit wisely.

For each statement below, mark whether you think it's True (T) or False (F):

a. A credit score ranges from 0 to 500.
b. Paying your bills on time is the most important factor in building good credit.
c. Checking your credit report can lower your credit score.
d. Using too much of your available credit can hurt your score.
e. You don't need to worry about your credit score until you're ready to buy a house.
f. Applying for several credit cards in a short time will improve your credit score.
g. If you have a high-paying job, you automatically have a high credit score.
h. There's nothing you can do if your credit report has an error.
i. Closing old credit cards is a smart way to boost your credit score.
j. Paying just the minimum on your credit card each month helps build good credit.

After completing the Credit Health Checklist, take a moment to evaluate your results. If you answered most of the questions correctly, congratulations! You're on the right track to understanding how credit works and how to manage it responsibly. Keep up the good work! However, if many of your answers were incorrect, it's a clear sign to revisit this chapter and dive deeper into the essentials of credit and debt management. Remember—the sooner you start mastering these concepts, the better prepared you'll be to make smart financial decisions as you enter adulthood.

Answers: a. False; b. True; 3. c. False; d. True; e. False; f. False; g. False; h. False; i. False; j. True

WHAT'S NEXT

Once you've tackled your debt, it's time to shift your focus from paying it down to building your net worth. By the time you've paid off your debt, the habits and discipline you developed will be well established. In the next chapter, you'll stop chasing money and start learning how to make money work for you through smart investing and building a portfolio that stays strong no matter what the market throws your way.

BREAKING THE TABOO

"More people should learn to tell their dollars where to go instead of asking them where they went."

— ROGER BABSON

Even in today's world, which, in many ways, is more open and transparent than it's ever been, it's still taboo to talk about money. Nevertheless, it's often associated with success, and when this is combined with social media, it's easy to gain a false view of the world and how finances work within it. This is a problem. Because we don't talk about the realities of money and practical financial literacy skills aren't taught in schools, many young people enter the adult world with a limited understanding. At the same time, they're watching online influencers who seem to have it all figured out, and the only logical conclusion is that they're just not that lucky. Very few people realize how important their mindset is and how much control they could have if they had good financial literacy skills.

The goal of this book is to help you realize your own power and give you the skills you've never been taught. This is the key to financial success, and that's something everyone has a chance at. If these skills were taught formally, there'd be no need for a book like this, but because they're not, it's important for it to reach as many young people as possible. You probably have a friend or two who always seem to be struggling with money—recommend this book to them and help them to see that there's something they can do about this. There's no reason for us not to talk about money. We can break the taboo, and this will benefit all of us.

You can play an important role in this change by contributing to the conversation online. This doesn't mean you have to start a forum or lose a lot of time to it. All you need to do to make a difference is leave a short review of this book. This way, you'll contribute to building a culture where it's ok to talk about money, and at the same time, you'll show new readers where they can find the education they've been missing out on.

By leaving a review of this book on Amazon, you'll show other teenagers that they have the power to shape their financial futures—and you'll show them exactly where they can find all the guidance they need to do it.

Reviews are a great way to get people talking and thinking, and they also help them find the resources they need. It's amazing how much of a difference a few sentences can make.

Thank you so much for your support. Together, we can empower the next generation with the financial knowledge that previous generations have lacked.

Scan the QR code below.

CHAPTER EIGHT

N—NET WORTH BUILDING

> *The best investment you can make is an investment in yourself. The more you learn, the more you'll earn.*

> — WARREN BUFFETT

Ever wonder why some people grow richer while others remain stuck living paycheck to paycheck? The secret lies in understanding and building net worth—the simple equation of everything you own minus everything you owe. Just like tracking your lap running times or your gaming stats helps to improve your performance, monitoring your net worth helps you to achieve financial goals.

WHAT IS NET WORTH?

By tracking your net worth, you can become more aware of what's working in your finances and what needs to change. When you track your net worth regularly, you're more likely to make smart financial moves. On the flip side, not knowing where you stand can leave you financially vulnerable (Financial Health Network, 2023).

How to Calculate Net Worth and What It Means

To calculate your net worth, start by listing your assets and liabilities. Assets are anything you own that has value—cash, investments, property, vehicles, collectibles, and even the potential value of your education and skills. Liabilities include everything you owe —student loans, credit card balances, mortgages, car loans, and unpaid bills. Subtract your total liabilities from your total assets. If the number is positive, you have a positive net worth, meaning your assets outnumber your debts. If it's negative, it means you owe more than you own. Unlike income alone, your net worth gives you a clearer picture of your overall financial health. Two people with the same salary can be in completely different financial situations depending on their net worth.

Knowing your current net worth gives you a solid starting point for financial planning. And don't stress—it's totally normal for a teen or a young adult to have a negative net worth at first due to student loans or car payments. Most people start out that way. What matters is the direction you're heading. Checking in on your net worth every few months helps you see whether your financial habits are getting you closer to your goals.

Building wealth doesn't start with a certain age or job title—it starts with consistent habits. Whether you've been saving for years or you're just getting started now, focusing on your net worth puts you ahead of those who are still waiting for "the right time." Knowing your net worth today helps you figure out where you're starting from and how far you have to go toward the kind of wealth you may not have thought was ever possible.

MASTERING THE BASICS OF INVESTING

When people hear the word "investing," many immediately picture Wall Street traders frantically buying and selling stocks, but real investing is much less dramatic and far more accessible. Investing means putting your money to work so it can grow over time. Instead of trading hours for dollars forever, you let your money work while you sleep. Think of investing as planting money seeds that grow into money trees that, in turn, produce more seeds. And just as different crops have varying growth rates, water requirements, and harvest seasons, different investments offer distinct risk profiles, growth potential, and time horizons. Think of various investment vehicles like cash investments, stocks, and real estate that combine to help investors create diversified "financial gardens" suited to their goals and risk tolerance.

Risk vs. Reward

The relationship between risk and reward is a fundamental investment principle. Generally, higher potential returns accompany higher risk levels—the possibility that actual outcomes differ from expectations. Risk manifests in various forms: market risk (overall market movements), inflation risk (purchasing power erosion), interest rate risk (price fluctuations as rates change), liquidity risk (difficulty converting to cash without loss), and specific risk (factors affecting individual investments). Rather than avoiding risk entirely, which creates its own risk of insufficient growth, successful investors manage risk through diversification, appropriate time horizons, and alignment with personal risk tolerance.

Within your finances, any income from your investment portfolio is considered as passive income. Passive income requires minimal ongoing effort once established, like royalties from creative work or dividends from investments. Income from investments typically derives from three primary sources, each contributing differently to overall income growth:

1. Capital appreciation occurs when assets increase in value—stock prices rising, real estate appreciating, or collectibles gaining worth.
2. Income generation provides regular cash flow—dividends from stocks, interest from bonds, or rent from properties.
3. Reinvestment compounds growth as these returns create additional earning capacity rather than being withdrawn for spending.

Different investment vehicles emphasize these components differently; growth stocks focus primarily on appreciation potential, while dividend stocks and bonds prioritize income generation.

Passive vs. Active Income

While active income might be your main focus now, gradually building a portfolio with passive income creates financial stability and, eventually, freedom. This income evolution typically progresses through distinct phases. Initially, active income dominates as you exchange time and skills for compensation—hourly wages, salaries, freelance payments, or business revenue requiring your direct involvement. With a consistent investment of surplus active income, portfolio income gradually grows—interest, dividends, capital gains, and rental income that requires management but not direct production work. Eventually, these investments can generate substantial passive income—regular returns requiring a minimal ongoing effort that supports your lifestyle regardless of whether you continue working. This progression from active to passive income represents the transition from working for money to having money work for you.

To get started investing, you need to understand that the most powerful force in investing isn't a stock-picking skill or market timing—it's time itself. Thanks to compound growth—when your earnings generate additional earnings, money grows exponentially over long periods. Consider this: If you invest $100 monthly starting at age 16, by age 65, you could have over $1 million, assuming a 10% average annual return. But start just 10 years later, at 26, and you'd need to invest nearly $300 each month to reach the same goal. Every year, you delay means missing out on compound growth's exponential power. So, even if you can only start with small amounts, starting early puts time squarely on your side. With simple interest, only your principal earns returns, like a plant that produces fruit without growing larger. But with compound interest, your returns generate their own returns—like a plant that grows larger each year, producing increasingly abundant harvests.

CREATING YOUR OWN PORTFOLIO

Your investment portfolio is essentially your financial garden—a collection of assets that grow at different rates and serve different purposes. Just as a garden benefits from variety, your financial portfolio grows stronger through diversification—spreading your money across different types of investments. The diversification principle extends beyond simply owning multiple investments; effective diversification requires including assets with different response patterns to economic conditions. During economic expansions, stocks typically outperform; during contractions, bonds often provide stability. When inflation rises, commodities and real estate can offer protection while bonds suffer. International investments sometimes thrive when domestic markets struggle. This non-correlation—assets moving independently or counter to one another—reduces overall portfolio volatility without necessarily sacrificing returns. Rather than attempting to predict which asset class will perform best (market timing), diversification acknowledges our inability to consistently forecast market movements and instead creates resilience across various scenarios.

Diversification

Combining assets with different risk-return characteristics can create portfolios with better risk-adjusted performance than individual components alone. Effective diversification considers correlations between assets rather than simply collecting different investments. For example, owning stocks in different technology companies provides less diversification than owning stocks across technology, healthcare, finance, and consumer goods sectors because technology stocks tend to move together.

As you begin investing, focus on understanding rather than complexity. Rather than trying to master all investment types at once, choose one that interests you, learn its fundamentals, and gradually expand your knowledge and portfolio. The following are a few popular types of investments:

- **Index funds:** Index funds warrant particular attention for beginning investors. These passive investments track market benchmarks like the S&P 500 (representing 500 large U.S. companies) or the Total Stock Market Index (representing the entire U.S. equity market). Their fundamental advantages include diversification (immediate exposure to many companies), low costs (minimal management fees), tax efficiency (infrequent trading generates fewer taxable events), and performance (historically outperforming most actively managed funds over long periods). Popular implementations include mutual funds (purchased directly from investment companies with minimum investment requirements) and exchange-traded funds, or ETFs (traded like stocks with no minimum investment beyond share price). Warren Buffett, consistently ranked among the world's wealthiest individuals, has recommended low-cost index funds as the most sensible investment choice for most people.

- **Individual stocks:** Playing the stock market allows investors to become part-owners of specific companies they understand and believe will grow over time. While potentially offering higher returns than diversified funds, individual stocks carry concentrated risk—company-specific problems can cause significant losses regardless of overall market performance. Successful stock investors typically develop expertise in specific industries, read company financial reports, follow management teams, and understand competitive dynamics. Beginners should approach individual stocks cautiously, perhaps allocating a small portfolio percentage as a learning opportunity while keeping most investments diversified. Stock investments require either self-directed research or professional guidance, along with emotional discipline to avoid reactive buying or selling based on short-term price movements.

- **Bonds:** These represent loans to governments or corporations in return for regular interest payments and eventual principal return. Their primary role in portfolios is providing stable income and reducing volatility, as they typically fluctuate less than stocks and often move in opposite directions during market stress. Bond characteristics include issuer quality (from ultra-safe Treasury bonds to risky "junk" bonds), maturity length (from short-term to 30+ years), and interest rate type (fixed or variable). Bond prices move in the opposite direction of interest rates—when rates rise, the value of existing bonds decreases; when rates fall, bond values increase. Young investors generally allocate less to bonds than stocks due to their lower growth potential, but including some high-quality bonds can reduce overall portfolio volatility while providing income.

- **Real estate investments:** Investing in real estate offers potential benefits, including income generation (rental payments), appreciation (property value increases), leverage (using mortgage financing to control valuable assets with limited capital), tax advantages (depreciation deductions and preferential capital gains treatment), and inflation protection (rent and property values typically rise with inflation). Direct property ownership requires substantial capital, management responsibilities, and reduced liquidity, making it challenging for beginners. Alternative access points include real estate investment trusts (REITs)—companies owning or financing income-producing properties that trade like stocks—and crowdfunding platforms allowing fractional property investments. These options provide real estate exposure with lower capital requirements and management burdens than direct ownership.

- **Entrepreneurial ventures:** These represent another investment avenue, allocating capital, time, and skills to build businesses generating both income and equity value. These investments range from side hustles requiring minimal startup capital to growth-oriented ventures seeking external funding. Successful business investments typically leverage personal expertise and market opportunities, potentially offering returns significantly exceeding traditional financial assets. However, they also carry concentrated risk and require substantial time commitment. For many investors, entrepreneurial pursuits complement rather than replace traditional portfolio investments, with business proceeds eventually funding more diversified holdings.

BUILDING WEALTH THROUGH INVESTING

When it comes to building wealth and your portfolio, making wise choices and maintaining habits that lead to financial success are more important than chance or winning the lottery. Be careful with your money, find other ways to make money, and put your money to work for you by investing it if you want to build wealth.

Assets vs. Liabilities

Knowing the distinction between assets and liabilities is the first and most important step. You may increase your wealth by owning assets such as a rental car, dividend-paying stocks, or a small company. Conversely, liabilities are items that deduct from your income, such as outstanding debts, vehicle loans, or valuable but quickly depreciating assets, such as electronics or high-end cloth-ing. To amass riches, one must put their assets ahead of their responsibilities. Consider putting that extra income into stocks or a side job instead of a new phone—both of which might end up paying for themselves.

Having several sources of income is another strategy used by the rich. Don't try to pay for everything with your allowance or the money you get from a side job. Consider other means of generating income. Some examples of internet businesses include YouTube channels, online stores, and marketplaces like Etsy. You can earn some more money by doing simple things like investing in stocks or selling your old clothing on eBay. The goal is to diversify your income streams so that you can continue to produce money even if one source of income slows down.

The last—and maybe most important—thing that rich individuals do is to reinvest their earnings. What this implies is that you should put your money to work for you rather than squander it immediately after receiving it. When you have extra money, rather than spending it on something you don't need, put it to work for you by investing it in stocks or even just purchasing and reselling stuff. The more time passes, the more money begins to earn interest and expands. It's like a snowball: It starts small, but as more snow is added, it becomes larger and rolls down the hill quicker.

In a nutshell, if you want to amass wealth, you should prioritize revenue-generating purchases, diversify your income streams, and continually reinvest your earnings. If you follow these steps, your money will increase at a rate you never imagined. And believe me —it will be worthwhile when financial worries no longer constrain you.

WHAT'S NEXT

After you've successfully made your money work for you, the next step is to focus on how well your digital financial transactions and data are protected, including daily transactions, banking data, and investments stored online. The following chapter will explore safeguarding your finances online and avoiding common scams.

CHAPTER NINE

E—E-SECURITY

An ounce of prevention is worth a pound of cure.

— BENJAMIN FRANKLIN

Picture this: You've worked hard to save money, build credit, and even start investing... then BOOM!—one scam email or password leak wipes it all out. Scary, right? But it's totally avoidable. The financial ecosystem has changed dramatically over the past decade. Paper statements and in-person banking have largely given way to mobile apps, digital wallets, and online investment platforms. For teens today, this digital-first approach to money management is the norm rather than the exception. Nearly all teens have some form of digital financial presence, whether through a banking app, payment services like Venmo or CashApp, or online shopping accounts. Each of these digital touchpoints represents both convenience and potential vulnerability.

Consider how your financial identity exists online—your bank accounts, credit cards, investment portfolios, crypto wallets, shopping profiles, and payment apps. Each platform stores valuable data about your financial habits and holds access to your hard-earned money. Just as you wouldn't leave your physical wallet unattended in a public place, your digital financial information requires deliberate protection. The stakes are actually much higher online, as digital theft can happen from anywhere in the world, at any time, often without immediate detection.

The consequences of lax digital security can follow you for years. While a stolen physical wallet might cost you whatever cash was inside, a compromised financial account could affect your credit score, drain your savings, saddle you with fraudulent debt, or even impact your ability to apply for student loans or rent your first apartment. For teens just beginning to establish financial independence, protecting your digital presence means protecting your future options.

THE DIGITAL MONEY LANDSCAPE

When people discuss financial safety, they often focus exclusively on investment risks or spending habits, but as technology has advanced, cybersecurity has become equally crucial to protecting your wealth. E-security means you're protecting your social media accounts and safeguarding your entire financial future. If you're using banking apps, investment platforms, or even just shopping online, your financial well-being is directly connected to your digital habits. Think of cybersecurity as the invisible shield that protects every dollar you've worked to earn and invest.

The Importance of a Secure Password

When it comes to digital protection, you need more than simple technical knowledge; you need to develop awareness and good habits. Especially for teens and young adults who conduct most of their financial lives online, understanding digital threats is as essential as knowing how to budget. Think about the last time you created an account online. Did you use a unique, strong password or your go-to password that you use everywhere? Research shows that 81% of data breaches involve stolen, weak, and/or default passwords (Verizon, 2017). Good digital security habits might seem tedious at the moment, but they're far less painful than recovering from identity theft or financial fraud.

Safe digital practices don't develop without intentional effort. They require conscious decisions and, at times, learning from close calls. Think about your current online behaviors. Do you verify before clicking links in emails? Do you check that websites are secure before entering payment information? These small actions create your digital security profile. But, even if you've had security lapses in the past, implementing stronger practices now can significantly reduce your risk moving forward.

PROTECTING YOURSELF IN THE DIGITAL WORLD

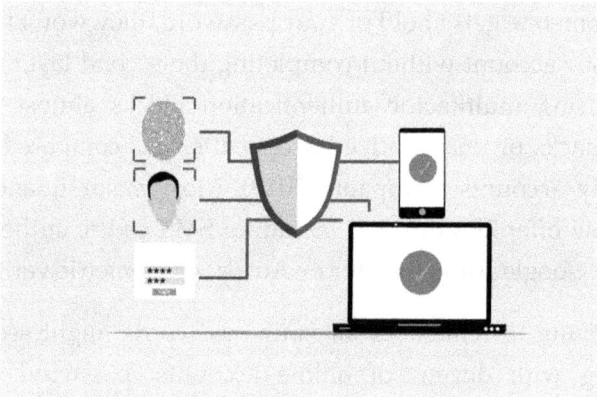

Your digital security toolkit doesn't require technical expertise—just consistent application of fundamental practices. The foundation of digital financial security begins with strong authentication—the process of proving you are who you claim to be when accessing your accounts. While passwords remain the most common authentication method, their effectiveness depends entirely on how you create and manage them. A study shows that more than 50% of people reuse the same password across multiple accounts, while 13% use the same password for all of their accounts (Google, 2019). This practice creates a dangerous domino effect: If one site experiences a data breach, attackers can potentially access all your accounts using the same compromised credentials. Instead of using variations of the same password, consider the passphrase approach —combining several random words into a memorable but difficult-to-crack sequence. For example, "correct-horse-battery-staple" is both easier to remember than "P@ssw0rd!" and exponentially more secure against automated cracking attempts. For financial accounts specifically, these passphrases should be unique and unrelated to phrases you use elsewhere.

Along with a strong password, two-factor authentication (2FA) provides an essential second layer of defense. When this is enabled, accessing your account requires both something you know (your password) and something you have (typically your smartphone). Even if someone gets ahold of your password, they won't be able to access your account without completing the second layer of verification. Using multifactor authentication blocks almost 100% of account hacks or unauthorized access attempts compared to password-only accounts (Cimpanu, 2019). Most major financial platforms now offer 2FA options, including SMS codes, authentication apps like Google Authenticator or Authy, or biometric verification.

While having to remember all your passwords might seem overwhelming with dozens of online accounts, password manager applications solve this problem by securely storing all of your unique passwords behind one master password. Think of it as a digital vault that remembers your complex passwords so you don't have to reuse the same one everywhere. A report showed that people using password managers experienced 61% fewer account compromises than those managing passwords manually (Google Security Blog, 2023). This simple tool eliminates the temptation to use easily guessable passwords or the dangerous practice of using identical passwords across multiple sites.

Besides the security of your account, the security of your devices themselves forms another critical protective layer. Each smartphone, tablet, and computer you use to access financial information represents a potential entry point for threats. Maintaining updated operating systems and applications is a critical security feature. Developers continually discover and patch vulnerabilities, but these fixes only protect you if you install the updates. Setting your devices to update automatically ensures you benefit from these security improvements without having to remember to check manually.

Beyond ensuring the security of your devices and accounts, you also need to take into account the network you're using. Public Wi-Fi networks—like those in coffee shops, libraries, or airports—often lack proper encryption, potentially exposing your sensitive data to anyone else on the same network. Avoid accessing your financial accounts on public networks whenever possible. If you must check your accounts while away from trusted networks, using a Virtual Private Network (VPN) creates an encrypted tunnel for your data, significantly reducing the risk of interception. Many VPN services offer affordable student or basic plans specifically designed for this kind of protection.

Keep a Healthy Dose of Skepticism

Beyond technical protections, developing a healthy skepticism toward unexpected communications protects your financial information from social engineering—attempts to manipulate you into revealing sensitive information. Before responding to emails or texts about your accounts, verify the communication through official channels. When creating security questions, choose answers that aren't easily researched through your social media. Keep in mind that legitimate financial institutions never request your complete password or PIN, especially through email or phone. These habits might seem paranoid until you realize that social engineering, not technical hacking, is the entry point for 70–90% of all data breaches (Cisco Security Report, 2023).

The concept of "digital hygiene" encompasses all of these security practices as regular habits rather than one-time actions. Just as you brush your teeth every day for long-term health benefits, consistent application of digital security measures provides ongoing protection for your financial life. Good digital hygiene includes regularly reviewing your account statements, periodically changing sensitive passwords, checking your credit report for unauthorized accounts, and staying informed about new security threats and protections. These routine practices help ensure that small vulnerabilities don't develop into major problems over time.

STEERING CLEAR OF SCAMS

Financial scams continue evolving, but they rely on predictable psychological triggers: urgency, authority, scarcity, and social proof. Understanding the psychology behind scams is your first line of defense. Scammers don't rely on technical sophistication as much as they exploit fundamental human tendencies and emotions. The feeling of urgency—that you must act immediately or lose an opportunity—bypasses rational decision-making processes. The authority scam uses your natural tendency to comply with official-seeming requests. Scarcity tactics create the impression that you'll miss out if you don't act fast. And social proof manipulates your inclination to follow what others appear to be doing. Recognizing these psychological triggers helps you pause and evaluate situations more objectively.

Phishing

Phishing remains the most prevalent form of financial scam targeting teens and young adults. These deceptive communications have evolved far beyond the obvious "Nigerian prince" emails of the past. Modern phishing attempts often perfectly replicate legitimate communications from financial institutions, complete with correct logos, formatting, and messaging tone. The only differences might be subtle—a slightly misspelled domain name, an unusual sender address, or requests that legitimate companies would never make. Training yourself to look for these small inconsistencies can prevent costly mistakes.

Scams

For teens entering the workforce, employment scams represent a particular danger. The promise of easy money, flexible hours, or work-from-home opportunities can be especially appealing when you're trying to balance school and income. However, legitimate employers will never require you to pay upfront for training, equipment, or background checks. They also won't send you checks to deposit before you've started working or ask you to purchase gift cards as part of your duties. Any job that pays far above the market rate for minimal qualifications deserves extra scrutiny. Remember that with legitimate employment, the money flows to you, not from you.

The rise of cryptocurrency and alternative investments has created fertile ground for investment scams targeting young people. These schemes capitalize on the fear of missing out (FOMO) and the complex nature of new financial technologies. They typically promise guaranteed high returns with minimal risk—a combination that simply doesn't exist in legitimate investing. Pressure to invest quickly "before the opportunity disappears" is another warning sign. Legitimate investment opportunities will provide clear information about risks, allow time for due diligence, and never guarantee specific returns.

Red flags of potential scams include artificial urgency ("Act now or lose this opportunity!"), requests for personal information through nonsecure channels, unfamiliar senders with slight variations in email addresses or domain names, grammar or spelling errors in supposedly professional communications, and offers that sound too good to be true. Trust your instincts—if something feels suspicious, take time to verify independently before taking action.

Identity Theft

Identity theft represents perhaps the most damaging category of financial fraud for young people just establishing their financial identities. When criminals get ahold of your personal information —like your Social Security number, birth date, or account numbers —they can open new accounts, take out loans, or make purchases using your identity. The damage is likely to take years to repair and significantly affect your credit score just as you're starting to build it. Limiting the personal information you share online, regularly checking your credit report (available free annually from each major credit bureau), and considering a credit freeze if you're not actively applying for new accounts can help protect you against identity theft.

What to Do if You Suspect a Scam

If you suspect you've encountered a scam, don't engage further. Don't click links, download attachments, or respond to the communication. Instead, contact the supposed sender directly through their official website or phone number (not the contact information provided in the suspicious message). Report potential scams to the

Federal Trade Commission (FTC), your state attorney general's office, and relevant platforms like your email provider. If you've already shared financial information, contact your financial institutions immediately to secure your accounts and consider placing a fraud alert or credit freeze with credit bureaus. These protective steps can prevent a momentary lapse from becoming a financial disaster.

The aftermath of falling victim to a scam requires prompt action. Every minute counts when unauthorized access occurs. Most financial institutions have 24/7 fraud departments specifically for these situations. Putting these contact numbers in your phone ahead of time can save crucial minutes during a security emergency. Document all communications with financial institutions and fraud departments, including representative names, case numbers, and promised actions. This documentation is extremely valuable if you need to contest charges or show that you reported problems promptly.

Implementing Cyber Security

Implementing comprehensive cybersecurity practices may seem daunting, but breaking it down into manageable steps makes it achievable.

- Start by securing your most sensitive financial accounts first—typically your primary bank account, credit cards, and any investment platforms. For these critical accounts, implement the strongest available security options, including complex unique passwords, two-factor authentication, and transaction alerts for any activity. Once these priority accounts are secured, gradually work through your other financial accounts using the same approach.

- Regular monitoring forms a crucial component of financial cybersecurity. Many financial institutions offer customizable alerts that notify you immediately when specific activities occur—large transactions, address changes, password resets, or unusual login locations. These real-time notifications serve as an early warning system for potential unauthorized access. Additionally, scheduling a weekly "financial security check" to review recent transactions takes just minutes but can identify problems before they escalate. Mark this recurring appointment in your calendar just as you would any other important commitment.

- The security of your mobile devices deserves special attention since they've become the primary access point for financial services. Beyond strong passwords or biometric authentication for the device itself, consider using app-specific passcodes for financial applications—creating an additional security layer if someone gains access to your unlocked phone. Review the permissions granted to apps, especially location access and photo/file access, limiting these to only what's necessary for functionality. And consider enabling remote tracking and wiping capabilities for your devices, allowing you to locate and secure lost or stolen devices before they can be compromised.

- Keep your devices updated with the latest security patches, enable biometric authentication when available, and set up remote tracking capabilities in case of loss or theft. Back up important financial documents to encrypted storage and develop a response plan for potential security breaches, including knowing how to contact your institution's fraud departments. These cybersecurity practices aren't just technical steps—they're essential safeguards for your financial future in an increasingly digital world.

- Developing a personal data backup strategy protects your important financial information from both security breaches and technical failures. Store digital copies of important financial documents (tax information, financial account numbers, identity documents) in encrypted cloud storage or on an encrypted external drive—never on unprotected devices or standard cloud services. Physical copies of critical documents should be stored in a secure location, such as a fireproof box or safe. This redundancy ensures that even in worst-case scenarios, you maintain access to your essential financial information.

Building digital resilience also means preparing for potential security incidents before they occur. Create a simple response plan that outlines immediate actions to take if you suspect unauthorized access to your accounts. This plan should include direct contact numbers for your financial institutions' fraud departments (not just general customer service), steps for placing credit freezes with the major bureaus, and templates for dispute letters. Having this information organized and accessible can significantly reduce response time during stressful security situations.

The landscape of financial cybersecurity continues evolving as both protection technologies and threats advance. Staying informed about emerging security risks and solutions doesn't require becoming a cybersecurity expert—just following reputable financial news sources or security blogs can provide timely awareness of major developments. Many financial institutions also offer educational resources about protecting your accounts, specifically within their systems. Taking advantage of these resources keeps your protection strategies current in a rapidly changing environment.

Finally, remember that financial cybersecurity isn't just an individual effort but a community responsibility. Share security best practices with family and friends, especially older relatives who may be less familiar with digital threats or younger siblings just beginning to use financial apps. Reporting scams to authorities helps protect others who might otherwise fall victim to the same schemes. This collective vigilance strengthens everyone's digital financial safety.

By implementing these comprehensive cybersecurity practices, you're not just protecting your money today—you're establishing security habits that will safeguard your financial well-being throughout your life. In a world where financial interactions increasingly occur digitally, these skills represent some of the most valuable financial education you can acquire. The investment of time and attention in security practices pays dividends through avoided losses and peace of mind about your financial future.

Interactive Element

Being new to digital financial transactions can make you more vulnerable to scams and fraud—especially since scammers are always finding new ways to stay ahead with the latest tricks and tech. To protect yourself, you need to be extra careful. Here are some red flags to watch out for in any form of communication to help you avoid scams and fraud:

- Be cautious when you receive an unexpected email, text, call, or message from someone you don't know or weren't expecting to hear from.
- Scammers often try to pressure you into acting immediately, using phrases like "limited-time offer," "urgent," or even threats like "your account will be locked."
- They may request sensitive information like your passwords, bank account numbers, credit card details, or

Social Security number. Legitimate companies don't ask for this out of the blue.

- They often push for payment through hard-to-trace methods like gift cards, wire transfers, or cryptocurrency.
- Look out for spelling mistakes, typos, or unprofessional language—it's a common sign of a scam.
- Avoid clicking on links or attachments in messages from unfamiliar sources. They could install malware on your device or redirect you to fake websites designed to steal your information.

WHAT'S NEXT

As you wrap up this chapter, take a moment to step back and look at the bigger picture. You've picked up a lot of important tools and knowledge around personal finance, and that's a huge win. While a year might feel far away, financially speaking, it's not too early to start planning. The choices you make today can help set you up for smarter decisions down the road. In the next chapter, you'll learn how to create a plan not just for the year ahead, but also for the next 3, 5, and even 10 years—helping you take solid steps toward your ultimate goals.

CHAPTER TEN
Y —YOUR FUTURE PLAN

" *A goal without a plan is just a wish.*

— ANTOINE DE SAINT-EXUPÉRY

You've learned how to budget, save, invest, spend wisely, and protect your money—but how do you tie it all together? The answer is a yearly plan. This is your big-picture roadmap for making all your financial dreams come true. Whether it's saving for a summer trip, buying a car, or even preparing for college, your yearly plan will keep you organized and focused so you can crush your goals, whatever they might be.

Although we don't know how things will all work out, what seems like an endless expanse into the future offers boundless opportunities and threats. One thing is clear: If the march of time has something in store for all human beings, the decisions you make today will determine whatever path is ahead for your life. "Your Future Plan" is not simply a map of the road but rather an open invitation to shape your destiny, set your own criteria for success, and create the life you desire.

When it comes to planning for the future, you can't just cross your fingers and hope for the best. Planning demands thoughtful deliberation, quick action, and steadfast dedication toward your goals. By applying the practical tips, you've learned so far, you're well on your way to creating a future that reflects your true strengths and desires.

You are in control of your own destiny; your future doesn't just happen. Now is the time to do something about that, whether you are just starting out on your career, beginning a new phase of life, or reassessing the path you are on. Your future depends on the choices you make today. Are you ready to start planning your future? Your answer to this question is the first step toward creating the life you've always dreamed of.

SETTING MILESTONES THAT MATTER

For teens like you, planning your future might sound like one of those things grown-ups talk about when they want to use the word "serious" with you. But planning your future does not have to be dull or stress-inducing—it's actually thrilling stuff! Just imagine you're designing your own adventure game where you get to choose the quests, characters, atmosphere, and final destination. And the best part? You are the star.

You don't need a perfect plan or a five-year spreadsheet. What you need is a dream. And that's where it starts. A spark. An idea. Something that makes your chest feel full when you think about it. Perhaps you could buy a coffee shop, write your own comic serials, or create apps to improve people's lives. What it is doesn't matter. Your dream is sound, even if it seems outrageous, arbitrary, or you're just the exception. Dreams are personal. No one can make them for you—not your friends, not your parents, not society. When you are excited about something, that is your clue; that is your starting point. Sometimes, you might not yet have one big dream, and that's perfectly fine, too. Just start with what you love doing. Start with curiosity. Your interests are like little breadcrumbs leading you toward something bigger.

Turning Dreams Into Reality

Now, here's the mind-bender: Dreams are great, but they're only the beginning. If you want to bring your dreams into reality, you have to turn them into targets. Dreams float around in your head, but target them and bring them down to earth so you can work on them. Let's say you dream of being a content creator. A good target might be, "I'll post a video every week for the next three months and learn to edit better." That's a lot more practical than just saying, "I want to go viral." You are adding structure to your dream. You are giving yourself direction. And when your targets are clear, a purpose starts to enter your behavior. That's when the magic starts.

But don't worry—you don't have to know every step along the way. It's not like following GPS instructions. In fact, hardly anyone does. Start your adventure by marking out a general destination, packing your bags, and navigating your way there. If you're dreaming of becoming a writer, your first goal might be setting up a blog. Every little bit helps—the dream you have in your head is created little by little. Slowly but surely you're building up momentum. Also, make sure your goals are your own, not what somebody else thinks is "realistic" and "respectable," because this is your life to live. So be sure that it's what you're really driven to do. If your heart's not in it, you won't be able to keep going. But if you truly care about your goals, even if they're difficult, you'll be more motivated to achieve them. Then, no matter how hard things get, you'll push right through it.

Also, give yourself room to switch directions. You don't have to be committed to just one dream for all time. Things change. Interests switch. Life throws you all kinds of curveballs. That's all a part of life. Maybe you start out wanting to be a vet, but in the end, it's environmental science or fashion design that catches your fancy. This doesn't mean failure; it means you're exploring, and that's thousands of times more important to your future happiness than anything else! Just keep moving. Keep on dreaming, keep planning, keep adjusting. Your future isn't something lying off in the distance waiting for you to come across by accident. It's yours to build, little by little, choice after choice, minute after minute. And you're not just waiting for anything good to happen—you're doing everything possible to make it happen!

MAPPING YOUR VISIONS INTO ACTION

You have a vision—a thrilling idea that makes your heart race just thinking about it. And now, you've done something big by turning your dream into milestones. But here's the real question: What will you do with it? This is where most people get stuck. They hesitate. They overthink. They say things like, "When I have more time..." or "Once I get better at this..." Wanna hear the truth? There's no perfect moment. If you're waiting for the stars to magically align, you'll be waiting forever. Action is what creates momentum. You don't need to be 100% ready. You just need to begin.

Taking that first step is like hitting "play" on your favorite song—it's only then that the music actually starts. If your dream is to launch a podcast, you don't need fancy equipment or a high-end studio. Start from where you are. Jot down topics you're passionate about. Record voice memos on your phone. Mess around with free editing apps. Every small step brings your dream closer to life. It's about progress, not perfection.

Think of success as a staircase—you climb it one step at a time. It doesn't happen all at once. Sometimes, you miss a step. Sometimes, it feels like you're just going through the motions. But as long as you keep climbing, you're still moving up. Some days will feel amazing—you'll be in the zone. On other days, you'd rather scroll for hours than do anything productive. That's totally normal. The key is to keep going, even when it's hard.

Let's say you want to be a content creator. A daily vlog might turn into a weekly upload. Over time, you start getting more views, and your audience grows. Or maybe your dream is to become a pro gamer. Ten minutes of practice a day turns into several hours each week. Before long, you're ready to compete. See how it works? You don't have to do it all at once. Just keep showing up. Consistency is what gets you there.

Making Dreams Achievable

Now, grab a piece of paper and write down that big dream—the one that feels too massive to achieve. Make it the title at the top of the page. If it feels overwhelming, if your mind immediately says, *That's too big*, tell it to hush. Just go with it. Let's break down that big dream—both financial and personal—into manageable steps you can start on today. Work backward: What needs to happen in 10 years, 5 years, 1 year, 6 months, and even this month to get you there?

Here's an example of how to break down a big personal and financial dream:

Ultimate Goal: Own a successful business by age 25.

- **10-year goal (age 25):** Run a profitable business with a strong brand, a loyal customer base, an efficient team, and smooth operations.

- **5-year goal (age 20):** Launch and stabilize the business after doing thorough market research, setting up financial projections, and building a solid strategy. Secure funding through savings, loans, or investments. Start growing brand awareness and gather customer feedback.
- **2-year goal (age 17):** Develop a business idea based on your skills, interests, and market demand. Validate it by researching your audience and checking out the competition. Explore funding options.
- **1-year goal (age 16):** Focus on building skills—take courses, attend workshops, or learn independently. Start networking through events and online communities, and find potential mentors or collaborators.
- **Monthly goals:** Research potential business areas that match your interests and strengths. Draft a simple business model. Create a basic online presence—maybe a simple website or social media page. Test your product or service on a small scale.
- **Weekly goals:** Set aside a few hours each day for research. Learn business-related skills. Get familiar with the basics of business finance.

Now, what if you try and it doesn't work out? First off, congratulations! You took the leap. That alone is huge. Everyone hits roadblocks. Everyone doubts themselves at times. When that happens, don't stay silent. Talk to someone. Reach out to a teacher, a mentor, a friend—or even a creator online who inspires you. People have been exactly where you are. Getting help doesn't make you weak; it makes you wise. No one makes it alone. Behind every success is a support system, someone who encourages, advises, or simply listens when it matters most. Let people help you. Tell others what you're working toward—you never know who might support your dream or want to be part of it.

Now, let's talk about something real: mistakes and failure. Things won't always go your way, but that's not a reason to give up—it's part of the process. Every role model you admire has failed more times than you'd guess. The only time failure wins is when you stop trying. Instead of letting it defeat you, learn from it. What didn't work? What can you do differently? Try again. That's how you grow.

MAINTAINING PROGRESS AND FORMING HABITS

Once you've set a goal and put it into action, you've done more than most people ever do. However, the next huge challenge still lies ahead—keeping at it. I mean, who isn't willing to start something new? But sticking to that plan is a different matter. You might get tired and be unable to keep going after you have started. And, as always, life has a habit of messing things up. That's where habits would help you.

Habits are a bit like brushing your teeth. You don't need to stir up your emotions every day to do it—you just do it. Why? Because you've made it a habit. That's how you need to approach your goals and how to achieve them. You need to make your behavior seamless and organic to meet this standard. No, it doesn't have to mean a radical life change. You're just making little adjustments at first.

To form good habits, however, it's not necessary to be perfect. You're bound to mess up. A day might slip by. The days could become hard to get through. And asking yourself - why'd you start this anyway? It's natural. The key is that you get right back on where you got off-track. One bad day does not negate all the good ones. The true power comes from bouncing back when it gets tough. That's where you grow. One nifty little trick for keeping up the pace is associating your habit with something you already do.

Do you always scroll your phone before bed? Well, then, spend five minutes journaling right after you get in bed.

Most importantly, don't forget to congratulate yourself! Every time you reach a goal—no matter how small it may seem—reward yourself with something enjoyable. Share your success with someone who's behind you. Celebrations help keep the fire burning and remind you why you began in the first place.

Creating your future plan toward establishing a more stable and resilient financial life is far from the end of your journey. In fact, it's just the beginning—it's where you start putting your plans into action. Whenever you need a refresher, you can return to this book or revisit any chapter that helps guide you.

CONCLUSION

By now, you've probably realized something important: You have full control over your financial future. With all the knowledge and tools you've picked up here—setting your own goals, making a budget, adjusting to life changes, and even starting to invest—things that once seemed like "grown-up stuff" now feel totally doable. What once felt out of reach is actually something you can start doing right now. And the sooner you begin, the more prepared and confident you'll be as you move into adulthood. So, now it's time to take everything you've learned—your skills, resources, and mindset—and use them to your advantage.

But here's a little secret most people never tell you: Adulting is sometimes just plain boring. There will be many days when it's not fun, not exciting, and not even interesting. But that's part of the process. Growth is not always dazzling—it's the quiet, steady effort, those times when you show up even though you don't want to. That's the difference between dreamers and doers.

So, if you're caught in the grind and feel like nothing is changing, trust the process. You are changing. Your actions are setting your future. You're developing discipline, concentration, and strength of character. These are the real breakthroughs. Anyone can start something. But adhering to it—that's where the magic occurs.

And at some point, months or years from now, you will look back and be so glad you didn't give up. Then you'll realize those small daily decisions and habits led to huge results. You'll recognize that making mistakes is not the end of the world, and success is not necessarily about being perfect—it's about not giving up and trusting the process.

No one has all the answers, especially when you are young, and that's perfectly fine. The most important thing is that you start. Dream big, be wise, and get up every day to take one small step at a time. Do something, achieve this or that goal, don't give up. Then your future won't be this big, dark unknown, but something you have already begun shaping through the choices you make, any goal that captures your attention, and every time you decide to keep trying even though things look bleak. Tell yourself, *This belongs to me.* You have the power to make it awesome. Sure, things might not go your way all the time—you will fail, you will feel lost, and you will want to quit—but that is just part of the journey. Don't fear changing direction or growing in new ways. That isn't failing —it's changing for the better. Stay attuned to yourself, keep learning, and keep moving forward. Any move you take, regardless of its size, puts a footprint a little bit closer to the kind of life that you want to live. So start now. Not tomorrow, not when you "feel ready." Right now. You don't wait for the future—you create it! One dream, one strut, and one day at a time. You've got this!

USE YOUR KNOWLEDGE TO HELP OTHERS!

You're in control of your financial future, and this puts you in a great position to motivate others. Please take a moment to help more young readers find this book so that they, too, have what they need to build a firm financial foundation.

Simply by sharing your honest opinion of this book and a little about how it's changed your relationship with money, you'll show new readers that they have the power to take control, and you'll show them exactly where they can find all the help they need to do so.

LEAVE A REVIEW!

Thank you so much for your support. Keep dreaming big, and keep taking action. The future is yours.

Scan the QR code to leave your review.

REFERENCES

American Public Health Association. (2021, October 26). *The impacts of individual and household debt on health and well-being*. APHA website. https://www.apha.org/Policies-and-Advocacy/Public-Health-Policy-Statements/Policy-Database/2022/01/07/The-Impacts-of-Individual-and-Household-Debt-on-Health-and-Well-Being

Bennett, K. & Goldberg, M. (2019, October 2). *Survey: Rising ATM and overdraft fees leave consumers paying much more than they did 20 years ago*. Bankrate. https://www.bankrate.com/banking/checking/checking-account-survey/

Buffett, Warren. (2024). *Warren Buffett quote*. A–Z Quotes. https://www.azquotes.com/quote/849234#google_vignette

Buffett, Warren. (2024, March 22). *Warren Buffett: 6 best pieces of money advice for the middle class*. Yahoo Finance.https://finance.yahoo.com/news/warren-buffett-6-best-pieces-110138500.html

Chen, J. (2022, September 14). *Demand-pull inflation*. Investopedia. https://www.investopedia.com/terms/d/demandpullinflation.asp

CFPB (Consumer Financial Protection Bureau). (2014, March 25). *CFPB finds four out of five payday loans are rolled over or renewed*. Official Website of Consumer Financial Protection Bureau. https://www.consumerfinance.gov/about-us/newsroom/cfpb-finds-four-out-of-five-payday-loans-are-rolled-over-or-renewed/

Cimpanu, C. (2019, August 27). *Microsoft: Using multi-factor authentication blocks 99.9% of account hacks*. ZDNET website. https://www.zdnet.com/article/microsoft-using-multi-factor-authentication-blocks-99-9-of-account-hacks/?utm_source=chatgpt.com

Cisco. (2024). *Underprepared and overconfident companies tackle an evolving landscape 2024 Cisco Cybersecurity Readiness Index*. Cisco. https://newsroom.cisco.com/c/dam/r/newsroom/en/us/interactive/cybersecurity-readiness-index/documents/Cisco_Cybersecurity_Readiness_Index_FINAL.pdf

Costa, B. & Gardener, N. (2021). *How gen Z shops and pays*. OliverWyman. https://www.oliverwyman.com/our-expertise/insights/2021/sep/how-gen-z-shops-and-pays.html

de Saint-Exupéry, Antoine. (2020). *A quote by Antoine de Saint-Exupéry*. Goodreads. https://www.goodreads.com/quotes/87476-a-goal-without-a-plan-is-just-a-wish

Danes, S.M. & Dunrud, T. (1993). *Children and money: Teaching children money habits for life*. Minnesota Extension Service Publication, HE-FO-6116, University of Minnesota.

Dunn, Allison. November 22, 2023. "150 Quotes on Saving, Investing, and Financial Wellness." Deliberate Directions. Accessed May 12, 2025. https://deliberatedirec tions.com/quotes-financial-wellness/.

Financial Health Network. (2023). *2023 U.S. trends report: Rising financial vulnerability in America.* Financial Health Network. https://finhealthnetwork.org/wp-content/uploads/2023/09/2023-Pulse-U.S.-Trends-Report-Final.pdf?utm_source=

Ford, Henry. (2014, July 15). *Whether you think you can, or think you can't ... you're right.* CPA Practice Advisor. https://www.cpapracticeadvisor.com/2014/07/15/whether-you-think-you-can-or-think-you-cant-youre-right/15980/

Franklin, Benjamin. (2020, March 14). *An ounce of prevention is worth a pound of cure.* VOA News. https://learningenglish.voanews.com/a/an-ounce-of-prevention-is-worth-a-pound-of-cure-/5326585.html

GoHenry. (2022, December 16). *How to teach your teenager about budgeting.* GoHenry. https://www.gohenry.com/us/blog/financial-education/how-to-teach-your-teenager-about-budgeting

Google. (2019). *There's still room for more education about online threats & security tools Security habits by generation 87.* Google Services. https://services.google.com/fh/files/blogs/google_security_infographic.pdf

Hallock, Addison H. (2009, October 23). *Quotable: Addison H. Hallock.* Military. https://www.military.com/paycheck-chronicles/2009/10/23/quotable-addison-h-hallock

Holt, F. (2021, July 9). *The invention of the first coinage in ancient Lydia.* World History Encyclopedia. https://www.worldhistory.org/article/1793/the-invention-of-the-first-coinage-in-ancient-lydi/

Internal Revenue Service. (2024). *Publication 501, dependents, standard deduction, and filing information.* Internal Revenue Service. https://www.irs.gov/pub/irs-pdf/p501.pdf.

Klontz, B., Klontz, T. & Kahler, R. (2010). *Wired for Wealth.* Simon and Schuster.

Majewski, T. (2015, November 5). *A brief history of Etsy on its 10th anniversary.* Built In NYC. Built https://www.builtinnyc.com/articles/brief-history-etsy

Mortimer, J. T. (2009). *The benefits and risks of adolescent employment.* National Library of Medicine, *17*(2), 8. https://pmc.ncbi.nlm.nih.gov/articles/PMC2936460/

Munger, T. T. (n.d.). *Thornton T. Munger Quote.* QuoteFancy. https://quotefancy.com/quote/1566718/Thornton-T-Munger-The-habit-of-saving-is-itself-an-educa tion-it-fosters-every-virtue

The Numbers. (2023). *Movie market summary 1995 to 2023.* (2023). The Numbers. https://www.the-numbers.com/market/

Nyrhinen, J., Lonka, K., Sirola, A., Ranta, M. & Wilska, T. (2023). Young adults' online shopping addiction: The role of self-regulation and smartphone use. *International Journal of Consumer Studies, 47*(5). https://doi.org/10.1111/ijcs.12961

Ozimek, A. (2021). *Freelance forward economist report*. Upwork.com. https://www.upwork.com/research/freelance-forward-2021

Parker, K., Graf, N. & Igielnik, R. (2019, January 17). *Generation Z looks a lot like millennials on key social and political issues.* Pew Research Center. https://www.pewresearch.org/social-trends/2019/01/17/generation-z-looks-a-lot-like-millen nials-on-key-social-and-political-issues/

Phillips, S. & Sandstorm, K. L. (1990). *Parental attitudes toward youth work.* Sage Journals 22(2), 160–183. https://doi.org/10.1177/0044118x90022002003

Ramsey, Dave. (2025). *Dave Ramsey quote.* BrainyQuote. https://www.brainyquote.com/quotes/dave_ramsey_520303

Rogers, Will. (2019). *Will Roger's Quote.* BrainyQuote. https://www.brainyquote.com/quotes/will_rogers_167212

Rogers, Will. R. (2019, October 22). *These 15 motivational money quotes will inspire you to get rich or die trying.* Grooming Lounge. https://www.groominglounge.com/blogs/thelounge/these-15-motivational-money-quotes-will-inspire-you-to-get-rich-or-die-trying#:

Smith, L. (2024, August 21). *Are you as frugal as Warren Buffett?* Investopedia. https://www.investopedia.com/articles/financialcareers/10/buffett-frugal.asp

Staff, J. & Mortimer, J. T. (2007). *Educational and work strategies from adolescence to early adulthood: Consequences for educational attainment.* Social Forces, 85(3), 1169–1194. https://doi.org/10.1353/sof.2007.0057

Taneja, R. M. (2014). *Money attitude: The impact of socialization in India.* Conference: International Seminar on Inequality and an Attempt at Bridging Them: Past and Present. Galgotias University. https://www.researchgate.net/publication/262843489_Money_Attitude_Impact_of_Socialization

Unilever. (2024). *Ben & Jerry's.* Unilever. https://www.unilever.co.uk/brands/ice-cream/ben-jerrys/

Vogel, H. U. (2013). *Marco Polo was in China: New evidence from currencies, salts and revenues.* Leiden: Brill.

Verizon. (2017). *2017 data breach investigations report.* Verizon Enterprise. http://www.verizonenterprise.com/verizon-insights-lab/data-breach-digest/2017/

White, A. (2020, December 15). *Millennials and Gen Z are the most likely to use mobile banking apps—here's why, plus budgeting tips.* CNBC. https://www.cnbc.com/select/why-millennials-gen-z-use-mobile-banking-apps/

Zhao, H., Gao, X., Jiang, Y., Lin, Y., Zhu, J., Ding, S., ... Zhang, J. (2021). Radiocarbon-dating an early minting site: The emergence of standardised coinage in China. *Antiquity*, 95(383), 1161–1178. https://doi.org/10.15184/aqy.2021.94

IMAGES

Bolovtsova, Katrin. (2020. February 21). *Assorted stickers and a planner* [Image]. Pexels.

Finanzgorillas. (2024, September 6). *50-30-20 Rule, Finance* [Image]. Pixabay.

Hassan, Mohamed. (2021, November 29). *Business, Target, Achieve* [Image]. Pixabay.

Hassan, Mohamed. (2025, January 2). *Cybersecurity, Security, Authentication* [Image]. Pixabay.

Hassan, Mohamed. (2018, March 20). *Cash, Investment, Business* [Image]. Pixabay.

Hassan, Mohamed. (2021, August 25). *Online Shopping, Credit Card, Ecommerce* [Image]. Pixabay.

Hassan, Mohamed. (2021, September 7). *Tax, Law, Heavy* [Image]. Pixabay.

Hliznitsova, Kateryna. (2023, April 17). *A woman putting money into a white envelope* [Image]. Unsplash.

Nattanan23. (2017, August 30). *Clock Money Growth Grow Time* [Image]. Pixabay.

Perry, Katelyn. (2023, May 12). *A pile of post it notes sitting on top of a blue table* [Image]. Unsplash.

Rilsonav. (2016, July 8). *Debt Loan Credit* [Image]. Pixabay.

Saeng, Allison. (2023, February 21). *A stack of coins sitting on top of each other* [Image]. Unsplash.

Scottsdale Mint. (2021, October 6). *A group of gold bars sitting on top of a table* [Image]. Unsplash.

www.ingramcontent.com/pod-product-compliance
Lightning Source LLC
Chambersburg PA
CBHW071702210326
41597CB00017B/2299